AND THE ODDS ARE . . .

The odds of writing a best-selling book:

- Assuming you can get your book published, your odds of writing a *New York Times* best-seller are 220 to 1.
- The odds of writing a "national best-seller" are 54 to 1.

The odds of reaching the summit of Mount Everest:

- About 20 percent of paying clients make it to the top. Thus, your odds of reaching the summit are about 4 to 1.
- Your odds of dying in the attempt are about 40 to 1.

The odds of having your identity stolen:

- Complaints of identity theft grew to 161,189 in 2002. Your odds of having your identity stolen in the coming year are around 200 to 1.
- The good news: federal law caps your liability at $50 for a lost or stolen credit card.

Gregory Baer is the coauthor of *The Great Mutual Fund Trap: An Investment Recovery Plan*. He has served as assistant secretary of the treasury for financial institutions, and was formerly managing senior counsel at the Federal Reserve Board. A graduate of Harvard Law School, he is now a partner in a Washington, D.C., law firm.

Visit www.LifetheOdds.com

LIFE : THE ODDS

Gregory Baer

A PLUME BOOK

PLUME
Published by the Penguin Group
Penguin Group (USA) Inc., 375 Hudson Street, New York, New York 10014, U.S.A.
Penguin Books (Canada), 10 Alcorn Avenue, Toronto, Ontario, Canada M4V 3B2
(a division of Pearson Penguin Canada Inc.)
Penguin Books Ltd, 80 Strand, London WC2R 0RL, England
Penguin Ireland, 25 St Stephen's Green, Dublin 2, Ireland
(a division of Penguin Books Ltd)
Penguin Books (Australia), 250 Camberwell Road, Camberwell, Victoria 3124,
Australia (a division of Pearson Australia Group Pty Ltd)
Penguin Books India Pvt Ltd, 11 Community Centre, Panchsheel Park,
New Delhi – 110 017, India
Penguin Books (NZ), Cnr Airborne and Rosedale Roads, Albany, Auckland,
New Zealand (a division of Pearson New Zealand Ltd)
Penguin Books (South Africa) (Pty) Ltd, 24 Sturdee Avenue, Rosebank,
Johannesburg 2196, South Africa

Penguin Books Ltd, Registered Offices: 80 Strand, London WC2R 0RL, England

Published by Plume, a member of Penguin Group (USA) Inc.
Previously published in a Gotham Books edition.

First Plume Printing, November 2004
1 3 5 7 9 10 8 6 4 2

The Library of Congress has catalogued the Gotham Books edition as follows:
Baer, Gregory Arthur, 1962–
Life : the odds (and how to improve them) / by Gregory Baer.
p. cm.
ISBN 1-59240-033-7 (hc.)
ISBN 0-452-28594-1 (pbk.)
1. Life skills—United States—Miscellanea. 2. Quality of life—United States—Miscellanea. 3. Conduct of life—Miscellanea. I. Title.
HQ2039.U6B34 2003
646.7—dc21 2003054953

Printed in the United States of America

For Arthur J. Baer,

who lives on every page

CONTENTS

Part 3:
REALLY BAD STUFF

INTRODUCTION

"No tears, no hugs, no learning." These were the immortal words used by Larry David, the co-creator of *Seinfeld,* to describe the animating premise of the show—that when people turn on the television in the hopes of having a few laughs, the last thing they want to see is characters undergoing emotional, spiritual, or intellectual growth. There will be no danger of that in *Life: The Odds.*

Instead, the muse for *Life: The Odds* will be Michael, the *People* magazine writer portrayed by Jeff Goldblum in the big-screen classic *The Big Chill.* Asked about his job, he explains to his fellow dysfunctional baby boomers, "Any article for *People* can be no longer than the average person can read while taking a crap." Wise words indeed.

There is every incentive to introduce this book by explaining how tragically uneducated the public is about probabilities and odds, and how the goal is to advance our understanding by making learning fun. Such an introduction would allow the author to explain to his kids that really the book was written for *them,* and for all the other folks out there who need some help with this important subject.

Kids, Daddy wrote the book for the money.

Nonetheless, there remains some danger that you might learn something by reading *Life: The Odds.* Directly, you'll learn the odds of all manner of events that concern, interest, or amuse us—everything from surviving cancer to winning a Rhodes Scholarship to being possessed by Satan. You'll learn how to calculate whether it's safer to drive or fly on your next trip. More important, though, you might indirectly learn to appreciate how to value and weigh the odds of improbable events. So the next time you're wondering whether a smallpox vaccine makes sense for you, or whether the government should be spending more on counterterrorism or auto safety, or whether you wish to assume the rollover risk of an SUV, or whether it's really *that* strange that you and your brother-in-law have the same dentist, you can think more rationally about the odds.*

You will not find many chapters devoted to the subjects most of us readily associate with odds: lotteries, sports gambling, and casino gambling. First, with respect to lotteries, here's all you really need to know:

- The odds are lousy.
- Everyone knows the odds are lousy.
- People buy lottery tickets anyway because a lottery ticket is their only hope of getting rich, and hope is important to people.

There would inevitably be tears and learning involved, so there will be no lottery discussions here.

As for other types of gambling, there is no shortage of books on these subjects, and no desire to repeat them here. In any event, the odds are generally posted, and the house always wins. That said, two casino games—keno and blackjack—are

* Consider, for example, the odds that your name is Greg Baer, and that you have only met or heard of two other people on earth with that name, both of whom are writers—Greg Bear in the science fiction world and Greg Baer, M.D., in the self-help world. (Please note that the Greg Baer at *www.gregbaer.com* who's talking a lot about crying, hugging, and learning is *that* guy.) Consider the odds that you learn of Greg Baer, M.D., because he is signed by the same editor at the same publisher in the same week.

examined in detail. Keno, because the odds are horrendously bad, not posted in any meaningful way, and universally underestimated. Blackjack, because it's the only casino game where the player can control the odds, and that's very cool.

But generally, the focus of *Life: The Odds* will be outside of Las Vegas and Monte Carlo—on the world where most of us live, where there are clocks on the walls and we have to pay for our own drinks. Here, we worry whether a large asteroid could be hurtling toward Earth; we wonder if there's any chance of bowling a perfect game during next week's league night; and we dream of one day climbing Everest or exploring space or, at the very least, writing a best-selling book.

So let's boldly go forth on our little adventure.

A BRIEF NOTE ON TERMINOLOGY

Odds are expressed in the form of, for example, 3 to 1. If the odds are *against* a particular outcome, as they frequently are, then the first number (3) represents the chance of failure in relation to the total (4); the second number (1) represents the chance of success in relation to the total. So if the *odds* of an outcome are 3 to 1 against, then the *chance* of it occurring is 1 in 4, and the *probability* of it occurring is .25, or 25 percent. (Conversely, odds in favor equate to a 3 in 4 chance and a probability of .75, or 75 percent.) In this book, where the odds are almost exclusively against, we'll refer to the odds "of" an event, and assume the odds are against, unless otherwise noted. So, if you see that the odds of writing a *New York Times* best-seller are 220 to 1, then assume that the chances are 1 in 221, with a probability of .0045.

Note also that when we speak of the "odds" in a given bet in gambling, that does not equate to the probability of winning. Rather, odds in the gambling context are what the casino or racetrack will pay you in the event that you win. The house makes money precisely because the two are different.

PART 1

LOVE AND MARRIAGE

DATING A SUPERMODEL

Determining the odds of dating a supermodel begins with two vexing questions: how does one define a supermodel, and how many are there? Most experts (including a lot of very strange, lonely men who have established websites for this purpose) identify the factors that allow a regular model to claim supermodel status:

- an exclusive contract with a major beauty company like Estée Lauder (a position held by Elizabeth Hurley, and now by Carolyn Murphy)
- regular appearances in the *Sports Illustrated* swimsuit issue (the bikini-clad Elle McPherson, Paulina Porizkova, and Kathy Ireland)
- regular cover shots in the major fashion magazines
- a regular, prominent role in the Victoria's Secret catalogue (the very naughty Stephanie Seymour, Tyra Banks, and Heidi Klum)

By most estimates, there are approximately twenty-five supermodels operating at any given time. These are the models who invade the public's consciousness, are profiled in "behind the scenes" cable television shows, and give disastrous performances in movies. While race, hair color, height, and general "look" may vary, there are two constants among the crème de la crème of the modeling industry: they are all thin, and they are all vapid. (Of course, you would be vapid, too, if you had dropped out of school at age fifteen and spent all of your time in makeup rooms, airplanes, and nightclubs.) But for many men, anorexia and vapidity are a good place to start, so the search for *supermodelis ignoramus* is a common dream.

The Odds

What are the odds? On the most elemental level, let's assume there are twenty-five supermodels, all of whom appear to be heterosexual or at least bisexual, and also there are 110 million adult men in America (as of the 2000 census), and that each supermodel will "date" five different American men per year (some fewer, and some dramatically more), and that the average male will be searching for a supermodel for ten years. The odds then work out to around 88,000 to 1.

Improving the Odds

These odds are misleading, though, as they assume that all men are equally well positioned to date a supermodel. In fact, occupation and geography skew those odds wildly.

Supermodels tend to date only certain types of men. To determine which types, the author conducted a search of the established supermodel literature (the tabloids) and was able to track the dating habits of forty-four supermodels, past and present. The occupation of each boyfriend or husband was identified. Obviously, one-night stands were not reported or recorded, but it is a fair assumption that the cohort of one-night stands would have a similar distribution of occupations as the cohort of boyfriends and husbands.

Here are the results:

Supermodels' Significant Others

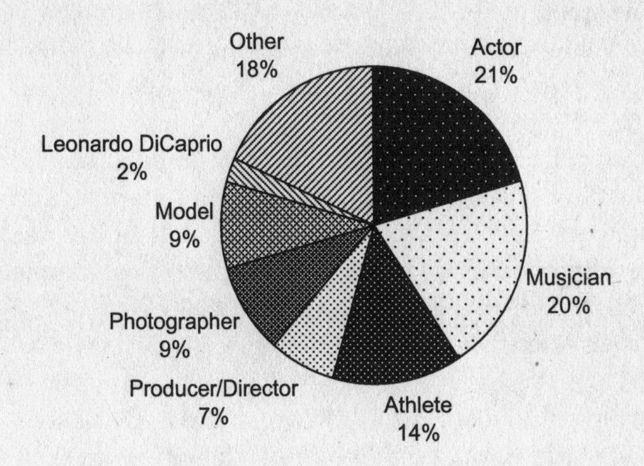

As the chart shows, actors, musicians, and athletes represent a majority of supermodel boyfriends and husbands. Leonardo DiCaprio himself represents a cottage industry in supermodel dating. But he is hardly the only actor to make the list—Mickey Rourke, Johnny Depp, Jason Patric, Sylvester Stallone, and even some guy named Norman Reedus have dipped their toes in supermodel waters over the last decade. Rock singers (Axl Rose, David Bowie, Rod Stewart, Ric Ocasek of the Cars) are not far behind.

Yet there is still hope for those who bothered to get a graduate degree. At least one lawyer and one doctor have broken through to the chamber of supermodel secrets. It also helps to put forth a little effort. One gentleman founded the Fashion Café—and ended up with supermodel Daniela Pestova. Photographers do well, as one might expect. Wealth doesn't hurt either. A "gallery owner" named Tim Jefferies has dated Claudia Schiffer and Elle McPherson, as well as porn star Koo Stark.

Occupation alone is not determinative, however. As with real estate, supermodel dating is heavily dependent on location, location, location. Practically every supermodel lives in one of two places: New York or Paris. You might have expected to see some California supermodels, but California is too far away from Europe, where most fashion shows are held.* So, if you aren't in one of the chosen industries and are looking to meet a supermodel, think about flying first class on Virgin Atlantic or one of the other transatlantic luxury airlines. (Surprisingly, given the travel required of supermodels, there is no record of any supermodel ever having dated a pilot.)

Adjusting our odds a bit, it's fair to say that if you are a star actor or the front man for a band based on the East Coast, then your odds are probably closer to 10 to 1. If you drive a truck in Minneapolis—well, keep on truckin'.

* California, however, is home to the pornography industry; while there are no hard data on the subject, the odds there presumably are better.

Being Poorly Endowed (or Romantically Linked to Someone Poorly Endowed)

Ask any man the three declarative sentences he most fears hearing, and the answer will always be the same:

1. "Bend over, spread your legs slightly, and cough."
2. "I think you're starting to lose some hair."
3. "You've got a small one."

Of these three, the last one is the harshest and most enduring. Even the clumsiest proctologic examination eventually comes to an end, and recedes from memory. For hair loss, there is the Hair Club for Men. But there is no Little Penis Club for Men; there is no cure; there is no hope.

To illustrate the importance of this subject, consider the three declarative sentences a man most *wants* to hear:

1. "My, that's a big one."
2. "My sorority sisters tell me you have a big one."
3. "Anna Kournikova told me you have a big one."

What is important, though, is not really absolute size, but size relative to other men. Some may remember a wonderful episode of the sitcom *Just Shoot Me*, where Finch (played by David Spade) is observed by his coworkers in a locker room. When they marvel at the size of his manhood, he expresses surprise, saying, "It doesn't look any bigger than the others I've seen."

"Seen where?" they ask.

"Well, you know, porno," he replies.

So, for our purposes here, the key questions are: What is average? What is noticeably above average? What is noticeably below average? and What are the odds of each? (We could go on to ask, Does it really matter? but let's not kid ourselves.)

The Odds

Determining average penis length (and girth) is a difficult undertaking. The first major study of penis size was conducted by researcher Alfred Kinsey in 1948. Kinsey determined that the mean length of an erect penis was 6.21 inches, with standard deviation of 0.77 inches. (A brief statististical refresher course: Standard deviation is a measure of how tightly grouped around the mean the data are. A low standard deviation means that all the results are about the same; a high standard deviation means that they vary greatly. In a normal distribution, one standard deviation—plus or minus—covers 68 percent of results, and two standard deviations—plus or minus—cover 95 percent of results.) Thus, according to the Kinsey data, 68 percent of penises were between 5.44 inches and 6.98 inches; 95 percent of penises were between 4.67 inches and 7.75 inches.)

Kinsey, however, can be justly accused of making the greatest possible mistake in penis-size studies: HE LET THE GUYS MEASURE THEMSELVES! Talk about book smart and common sense stupid. Try to picture some guy being handed a postcard (the Kinsey methodology) and asked to go off to a room to measure. (The private room was necessary because all measurements had to be at erection, which is not so easy with Dr. Kinsey staring at you.) Given that arrangement, do you think maybe subjects are going to be rounding up a little when they return to report results to Dr. Kinsey's cute little research assistant?* Not likely.

Now, here's an interesting thought: there is one group of people with a keen financial interest in making certain that an accurate gauge of penis size is obtained: condom manufacturers. Sure, for marketing reasons they're going to label each of their products, regardless of size, with some flattering title: the "Gargantuan," the "Monster," the "ICBM." But at the end of the day, they do need to get the size right.

Sadly, the major condom company studies used an even more flawed methodology than Dr. Kinsey's, and obtained similar results. Durex conducted an Internet survey and received responses from 2,936 men in seventeen countries. There was no effort to verify the data. Thus, this study suffered not only from the natural tendency to exaggerate in Dr. Kinsey's study** but also from a self-selection problem: poorly endowed men were certainly far less likely to be reporting penis size over the Internet than well-endowed men. (The opposite phenomenon would be expected if the survey only included ex-wives.) Not surprisingly, under the Durex

* In truth, the appearance of the Kinsey assistants is lost to history, but it's not hard to picture, especially in a 1940s nurse's hat with the front three buttons of the white blouse casually unbuttoned, but we digress.

** Actually, in the Durex study, with data being reported anonymously rather than to the hypothetical cute reasearch assistant, the incentive should have been to *understate* penis size. Why? Because when the aggregate data were reported, any given participant would have ended up looking relatively better if the reported mean size were *smaller*. Because men are stupid and shortsighted, however, we can be confident that no such analysis actually occurred, and that the subjects simply went with a "Mine is huge!" ethos.

methodology, average penis size "grew" to 6.4 inches, with a wider standard deviation of 1.1. (The wider standard deviation is consistent with the subjects' engaging in exaggeration.)

Most recently, though, the company Ansell (the folks who proudly bring you LifeStyles condoms) conducted a study where measurements were conducted by the researchers themselves. Each measurement was conducted by two nurses, supervised by a doctor. How do you arrange such a study? Well, you head down to Cancún, Mexico, during Spring Break and get a bunch of drunk guys to drop their drawers. This particular study included 401 men who happened to be at the Dady Rock nightclub. (The actual survey size was 300, though, since a rather surprising 101 of the participants were unable to produce the erection required for the measurement—an implicit tribute to the quality of the tequila at the Dady Rock, or welcome comfort for Bob Dole and all those other middle-aged Viagra users.)

The Ansell/LifeStyles study probably suffered from self-selection bias, as your better endowed Dady Rock patron was probably more likely to hop off his barstool and head to the LifeStyles tent. Still, at least the measurements were accurate. The result? Average penis size drooped to 5.88 inches, with a smaller standard deviation of 0.83 inches.

There is, of course, a more reliable methodology that has never been used: asking Madonna. But absent a comprehensive study of that sort, it's probably safe to assume that the average penis size is closer to 5.5 inches. That means that you have 2 to 1 odds of having a penis between 4.7 and 6.3 inches. You have 19 to 1 odds of having a penis between 3.9 inches and 7.1 inches. Beyond that, you're looking (staring) at a serious outlier.

AND NOW FOR THE LADIES. . .

We know Barbie's measurements. Adjusted to the size of an average woman, her measurements are 38-18-34. But is that *normal*? In common parlance, the ideal female measurements are 36-24-36. Is *that* normal? Well, according to the National Textile Center (the folks who do research to make sure clothes fit), the average female measurements are 38-32-40.

But we'll focus on the first number—the breasts. According to the fun folks at *www.sexualrecords.com*—well worth a visit if you want to know, say, average hymen thickness or average erection angle*—the average breast size is 35.9 inches, yielding an average bra size of 34B. Cup sizes break down as follows: A=15%; B=44%; C=28%; D=10%, with the rest (3%) outliers (Twiggy, Dolly). Breasts have been growing in recent years due to improved diet and the increased use of birth control pills.

Of course, women with tiny breasts have a surgical option unavailable to men with tiny penises. And big breasts are *in*! According to the American Society for Aesthetic Plastic Surgery (ASAPS), there were 216,754 breast augmentation surgeries performed in 2001. On the other hand, there were 114,926 breast *reductions* performed the same year, so the net number of breast-size increases is only 101,828. (Oddly, breast reductions are frequently covered by insurance, but breast augmentations are not.) Of course, breast augmentation is just the tip of the cosmetic surgery iceberg. According to ASAPS, there were 8,470,363 cosmetic procedures performed in 2001—a 48 percent increase from the previous year, and a 304 percent increase from the annual trend over the previous five years.

What explains the meteoric rise in the occurrence of

* 0.05 to 0.10 inches and 15 degrees above horizontal, respectively, in case you're curious.

plastic surgery? Three words: Botox, Botox, Botox. The newest and most popular cosmetic procedure is Botox injections, with over 1.6 million procedures. Trailing close behind are chemical peels (1.4 million) and collagen injections (1.1 million). Forget breast augmentation—cosmetic surgery has moved above the neck!

You may assume that your odds of opting for cosmetic surgery are larger if you do not live in Hollywood or the surrounding California area. You would be wrong. According to ASAPS, which analyzes statistics by region, the thirteen-state mountain and western region, which includes California, accounted for only 27.5 percent of procedures. The botox was flowing far faster in New England and the Mid-Atlantic, which account for 37.0 percent of all procedures.

MARRYING ROYALTY

The castle, the title, the money, the loyalty of one's devoted subjects—yes, it's nice being Michael Eisner, CEO of Disney. But Mr. Eisner is married, so if you want all those things and more, better odds lie in seeking out a royal marriage. And in the twenty-first century, with most European royalty long forgotten and Middle and Far Eastern royalty running for their lives, the only royalty really worth marrying into is English.

We'll have to define our terms here a bit, though. Strictly speaking, royalty connotes only those members of a royal family. In England, this means the Windsor clan—small in numbers, big in ears, except for those demigods William and Harry, who favor their mother and do not lack for attention. Given the limited options and stiff competition for the palatable ones, you'll need to broaden your search a bit to improve your chances.

The Odds

One's best odds come with seeking a member of the English peerage or baronetage. The concept of the peerage dates back to the fourteenth century and King Edward II, who kept a fixed list of those members of the landed gentry who were eligible to attend meetings of parliament. In past centuries, leaving one's castle for government service had been seen as an annoyance and imposition, but under Edward II it came to be seen as a privilege (akin to voting). Furthermore, when

Edward II and his successors decided that the right to attend meetings would be inherited, the honor of being a member of the peerage grew in importance.

Today, a "peer" refers specifically to certain persons who hold a title of honor (or, as the British would say, honour), including duke, marquess, earl, viscount, and baron. Peers qualify for membership in the House of Lords, with the House of Commons generally limited to commoners.*

While a peerage would certainly be ideal, a baronetage should also suffice. A baronetage is also a hereditary title of honor, but just short of a peerage. While membership in the House of Lords is not part of the package, other good things are. Besides, time saved traveling to London for governance is more time for shooting, horseback riding, and sumptuous banquets.

One should probably draw the line there, however. A knighthood is certainly impressive, but only for the holder, not the spouse. Knighthoods generally are conferred for the knighted one's lifetime only. (Think Sir Paul McCartney and Sir Sean Connery.) While a knight gets a title (the "Sir" or "Dame" part), the spouse gets nothing, and the spouse's children inherit no title. Plus, people who are routinely referred to as "Sir" start to get a little too big for their britches, and quickly cease taking out the garbage and putting down the toilet seat.

(As the toilet seat analogy suggests, analysis here is lim-

* This gets a bit complicated. For example, while peerage is generally synonymous with membership in the House of Lords, not every lord is a peer. Every peer, however, is a lord (except, perhaps, dukes, who are so significant that they are not generally referred to as lords, though they are in fact lords). But for marrying purposes, matters are relatively simple. You know from birth who the peers are. Those achieving lordship through less traditional means—for example, extraordinary achievement or high office—often do so too late in life to be a practical alternative. Plus, they tend to be poorer.

mited to the odds of a female finding a male peer or baron, as matters get even more complicated the other way around.)

Of course, while to some extent a peer is a peer, some peers are of higher rank than other peers (even though that sounds oxymoronic). So, if you get invited to a shooting party, and there are two guys standing by the bar, each looking equally dapper, it pays to know which one is of higher rank.

Here, the "order of precedence" should be your guide. For this purpose, "precedence" governs the order in which you are seated at dinner, listed in a roster of those attending a function, march at a royal funeral (with most important to the rear), and share in any number of other royal bounties. The order of precedence is not the same as the order of succession, since it includes persons ineligible to wear the crown, but there is considerable overlap at the top.

In the most senior ranks, the list does not discriminate by sex, but below the top twenty-nine peers, the order of precedence is determined by the man in any couple; women have a separate list, in case unaccompanied. So, for example, the sovereign is *numero uno* (seated and listed first, marching last). The sovereign's uncle's wife (currently, Princess Alice, Duchess of Gloucester), is 18th; the Archbishop of Canterbury is 30th; the lord Privy Seal (who knew that was a *person*?) is 38th; and Marquesses in the peerage of Ireland created before the Union of 1801 are 57th. The list hardly stops there, though, with Eldest sons of Knights of the Garter (too easy) seated 137th, and younger sons of Knights Bachelor 171st. And then it's time for the ladies. Lest you think it's all about royal blood, the Master of the Horse (if not a peer) comes in 90th, well ahead of the average Knight.

With these rules in mind, one's indispensable reference is *Burke's Peerage and Baronetage*, first published in 1826 and listing all the royal houses, from Abdy to Zouche. For one in a hurry—say, a lady has met a gentleman at a party, and just has time to pop out to her Land Rover for a quick peek at the copy of *Burke's* on the backseat—the latest edition contains an index listing all 120,000 living persons of noble heritage.

Obviously, your odds are going to be best living in the United Kingdom, where these folks—well, they're not really folks—tend to live. With roughly 60 million inhabitants, the odds of finding royalty there are around 500 to 1, with women having a much better shot than men. In the former colony, the United States, the overall population is larger, the peers are fewer, and the odds are far longer.

Improving the Odds

Those seeking to identify British royalty are fortunate indeed that the latest edition of *Burke's*, its 106th, was published relatively recently, in 1999. The 105th was published in 1970. So if you were, say, a Fox producer looking to cast *Who Wants to Marry an Earl?* circa 1998, you were stuck with either doing your own research or settling for some decidedly long-in-the-tooth contestants. Now, with the 106th edition, it's open season once again.

But if one is really set on finding royalty, one can expand beyond British royalty. You probably don't want to be marrying into the Shah of Iran's clan, but the latest *Burke's* has been expanded to include Scottish and Irish chiefs, and Scottish and feudal barons. There is even an edition now devoted to ancestors of the American presidents. So, today's social climber can look to multiple ladders.

MARRYING A MILLIONAIRE

There are many ways to turn yourself into a millionaire. Pursue and succeed at a high-income or stock option–rich occupation. Earn a steady salary, invest wisely, and live frugally. Or, of course, marry someone who has already done these things or, better yet, who simply inherited his or her millions and thus is not worn down by toil.

The Odds

The good news is that the number of millionaires is increasing. Estimates are that there will be 5.6 million millionaires in the United States by 2005. Furthermore, with more women joining the millionaire ranks, new options are opening up for male gold diggers.

Unfortunately, while there are thus more millionaires than ever before, very few of them are single. According to Thomas Stanley's *The Millionaire Mind*, 92 percent of those millionaires are married, 2 percent are divorced, 2 percent are single and have never been married, and 4 percent are widowed—meaning that 8 percent of our 5.6 million millionaires, or 448,000 millionaires, are single. As of the 2000 Census, of the 221 million Americans age fifteen and over (which is roughly marrying age in most states),* 60 million (7 percent)

* Most states allow marriage with parental consent at age sixteen or seventeen. You will be unsurprised to hear that the age falls to fourteen in Alabama, Texas, and Utah. But the shockers are Kansas (fourteen for males and twelve for females), Massachusetts (same), and New Hampshire (fourteen and thirteen).

had never been married, 22 million (10 percent) were divorced, and 15 million (7 percent) were widowed—a total of 97 million singles.

In other words, the odds that the person you're dating is a millionaire are 215 to 1.

Improving the Odds

From an analytical perspective, improving your odds of marrying a millionaire consists of four steps: (1) locating a millionaire; (2) attracting the millionaire; (3) dating the millionaire; and (4) getting the millionaire to the altar. But how to move from step (1) to step (4)?

For advice here, it's best to rely on the experts: scheming women who have used every possible artifice to date rich men and acquire their lifestyles. Thankfully, at least two such women have written books: Lisa Johnson, author of the comprehensive *How to Snare a Millionaire,* and Ruth Leslee Greene, author of the hard-hitting *How to Marry Money*. Qualified? You bet. Ms. Johnson is described on her book jacket as "a multimedia journalist and food critic" who "has dated many men of wealth, most of whom have proposed to her." Ms. Greene's biography is not mentioned anywhere in her book, and is not discoverable on-line; there is a good chance that the author's name is not even Ruth Leslee Greene. But is the fact that the book was written by a pseudonymous author a reason not to take it seriously? To the contrary, it may well have been written by a woman who wishes to hide her scheming secrets from her incredibly wealthy husband or his suspicious family. (Three words: Anna Nicole Smith.)

In any event, standing on the narrow shoulders of these social climbing giants, we proceed to think through the spotting, pursuit, and conquest of the rich.

Locating Your Millionaire

Obviously, some places are much better for spotting a millionaire than others. Among the hot spots that common sense and our experts' wise counsel identify as good places to explore are yacht clubs, Episcopal churches, polo matches, art auctions, and charity functions.

Equally important, though, is avoiding places where you are unlikely to ever run into a millionaire, and thus are wasting your time. Ms. Johnson, in *How to Snare a Millionaire,* contributes a comprehensive list, including:

- Laundromats
- Wal-Marts
- tractor pulls
- Denny's
- adult magazine shops
- professional wrestling matches
- cosmetology school
- auto parts stores

Of course, spotting one's quarry is in some ways the easiest step. The true skill comes in what follows.

Moving in for the Will

Approaching millionaires must be done carefully, as they are nervous creatures prone to flight. Generally, your approach will need to vary with the situation, but Ms. Greene, in *How to Marry Money,* offers some tried and true approaches:

- On a cruise, wait until your quarry leaves his deck chair to get a drink or use the facilities, and then occupy it. When he returns, first quarrel over the chair, then apologize, then flirt.

- Determine the brand and number of tennis balls your quarry uses. Buy identical balls, and occupy the court adjacent to his. Gradually, steal all of his, and insist they are yours. (While Ms. Greene is too sweet to describe the method as such, at this point one might note that you have him by the balls.) Then, according to Ms. Greene, "After being such a bitch, the very least you can do in apology is invite him for some refreshment."
- Buy a dog and walk it in wealthy neighborhoods. Teach the dog to bite whomever you designate, and point it toward preselected millionaires. In the course of offering apologies and first aid, strike up a conversation.

If you use these methods wisely, you have the chance to make that all-important first impression, and entice him to ask you out. Now the game's afoot!

Dating and Beyond

Once you have a millionaire smitten, the job is to keep him interested.

First date. The first date will be crucial. In *How to Snare a Millionaire*, Ms. Johnson offers comprehensive advice on what you should say and do on the first date, with the prime directive being to "establish common interests, and express great enthusiasm for them." Equally important, though, are things to avoid on that first date—for example, Ms. Johnson counsels against:

- insisting that your children come along
- asking him to help you out with your "financial difficulties"
- chewing tobacco

- expecting him to empty your garbage
- inquiring about the beneficiary of his life insurance policy

If you feel you may have a hard time navigating this minefield, you may wish to write these down, or pick up a copy of her book.

Inviting him into your lair. When your millionaire visits your apartment or house to pick you up for a first or subsequent date, you will need to present an attractive lifestyle. Here Ms. Johnson's book again provides some keen insights. You may wish to have on display art prints, an espresso machine, an inviting couch, and big fluffy towels. Items that Ms. Johnson counsels avoiding: fuzzy toilet seat covers; crossword puzzle books; an abundance of medication sitting around; and a "zoo" of stuffed animals. You will not wish to look eccentric or needy.

Closing the Deal

Once hooked, your millionaire needs to be brought on board promptly, without time for reflection. Ms. Greene counsels strongly for elopement, thereby minimizing the potential for familial disruption. Failing that, she offers this sage counsel:

> Try to tie the knot in a very small church or registry office so you won't have to invite your embarrassing cousins or old friends among whom your improved fortunes have triggered cupidity and envy. Detach from them without rancor, and adjust to the rarefied atmosphere of your new rich life. A small price, really, for such a large success!

Really, that says it all.

D-I-V-O-R-C-E

In this chapter, we'll quickly dispose of the tedious main topic—your odds of getting divorced—and then move on to the more interesting sidebar—the odds of celebrities getting divorced!

The Odds

As for boring old you, the odds are a bit in dispute. According to the National Center for Health Statistics, the odds are 1.3 to 1 that a first marriage will survive without separation or divorce for fifteen years. (Put another way, the probability of divorce or separation within fifteen years is 43 percent.) The study was based on the National Survey of Family Growth, a nationally representative sample of women age fifteen to forty-four in 1995.[1]

According to the Census Bureau, however, your odds are not as good. The Bureau believes that the odds of divorce for newly married couples are about even money, 1 to 1. Why the discrepancy? Determining divorce odds is actually very difficult, because divorce patterns vary significantly by generation, and there is always a significant lag in the data. Thus, it's difficult to predict how the present generation of couples will behave.

The most meaningful illustration of divorce rates is presented by the Census Bureau:

Percentage of First Marriages Reaching Stated Anniversary

Sex and year of marriage	Anniversary 5th	10th	15th	20th	25th	30th	35th	40th
Men								
1945–49	95%	91%	87%	83%	79%	76%	73%	71%
1950–54	97%	91%	85%	80%	77%	74%	71%	67%
1955–59	94%	86%	79%	74%	70%	66%	63%	NA
1960–64	94%	82%	74%	67%	73%	59%	NA	NA
1965–69	91%	76%	66%	61%	57%	NA	NA	NA
1970–74	88%	72%	63%	58%	NA	NA	NA	NA
1975–79	86%	71%	62%	NA	NA	NA	NA	NA
1980–84	87%	74%	NA	NA	NA	NA	NA	NA
1985–89	88%	NA	NA	NA	NA	NA	NA	NA
Women								
1945–49	95%	90%	86%	81%	75%	70%	65%	59%
1950–54	95%	89%	83%	77%	72%	66%	61%	56%
1955–59	95%	88%	79%	72%	66%	62%	58%	NA
1960–64	93%	82%	72%	64%	59%	55%	NA	NA
1965–69	90%	76%	65%	59%	44%	NA	NA	NA
1970–74	87%	71%	62%	56%	NA	NA	NA	NA
1975–79	85%	70%	61%	NA	NA	NA	NA	NA
1980–84	86%	73%	NA	NA	NA	NA	NA	NA
1985–89	86%	NA	NA	NA	NA	NA	NA	NA

Source: U.S. Census Bureau, Survey of Income and Program Participation (SIPP), 1996 Panel (marriages ending in death excluded from calculation)

As the chart shows, long-term divorce rates can be calculated, by definition, only for couples who were married a long time ago. So we can't calculate the odds of a marriage from the 1990s surviving a lifetime, or even twenty years, because we haven't had sufficient time to measure. The best we can do is calculate the five-year or ten-year odds; compare them to those odds from earlier times; determine if there has been an upward or downward trend; and then assume that any such trend will hold true over longer periods. Thus, what is significant about the chart is that for just about every anniversary that can be compared (with the fifth anniversary being the one that can be compared for the most five-year groups), the percentage of surviving marriages is in steady decline, with only a slight uptick for those married in 1980–84. Thus, for example, the percentages of male first marriages surviving until a fifth anniversary declined steadily from 95 percent to 88 percent between 1945–49 and 1985–89, and the percentage of marriages surviving until the twentieth anniversary declined from 83 percent to 58 percent between 1945–49 and 1970–74.

Thus, the pessimistic Census Bureau estimate for today's married couples is based on the assumption that the percentage of marriages surviving forty years likewise has declined appreciably since 1950–54 (the last available data set), when it stood at 67 percent. Declined all the way to even money odds.

Improving the Odds

There is no shortage of books providing advice on how one can have a successful marriage. Most of the leading indicators, though, are outside our power to control. For example, the fact that your parents were divorced is a particularly bad sign. But you can't divorce your parents (and, even if you could, that would probably be a leading indicator, too). And while numerous studies and books tell us that the greatest chance of divorce comes in the first seven years of marriage, that's hardly a great reason to remain together—"Honey, I know you despise me and are sleeping with my best friend, and I shouted obscenities at you on the Ricki Lake show, but statistically, if we could just hang on until the second quarter of next year, our odds of divorce will decline appreciably."

One surprisingly predictive indicator of divorce that you *can* control is geography. Divorce rates vary considerably based on the state where you live, as shown in the following table:

Divorces per One Thousand Residents

Rank	State	Rate	Rank	State	Rate
1	**Massachusetts**	2.4	27	Utah	4.7
2	**Connecticut**	2.8	28	**Delaware**	4.8
3	**New Jersey**	3.0	29	Montana	4.9
4	**Rhode Island**	3.2	30	Missouri	5.0
5	**New York**	3.3		West Virginia	5.0
	Pennsylvania	3.3	32	North Carolina	5.1
7	**Wisconsin**	3.4		Colorado	5.1
	North Dakota	3.4	34	Georgia	5.2
9	**Maryland**	3.5	35	**Oregon**	5.3
10	**Minnesota**	3.6	36	Texas	5.4
	Louisiana	3.6	37	Alaska	5.5
12	**Illinois**	3.7	38	**Washington**	5.6
13	**D.C.**	3.9	39	Mississippi	5.7
	Iowa	3.9	40	Kentucky	5.8
15	Nebraska	4.0		Arizona	5.8
	Vermont	4.0	42	Florida	5.9
17	**Michigan**	4.1	43	**New Mexico**	6.0
18	South Dakota	4.2	44	Idaho	6.2
	South Carolina	4.2		Alabama	6.2
	Hawaii	4.2	46	Indiana	6.4
21	**California**	4.3	47	Wyoming	6.5
22	**Maine**	4.4	48	Tennessee	6.6
	New Hampshire	4.4	49	Oklahoma	6.7
24	Ohio	4.5	50	Arkansas	7.1
25	Virginia	4.6	51	Nevada	9.0
26	Kansas	4.7			

Source: *www.divorcereform.org;* Centers for Disease Control and Prevention/National Center for Health Statistics.

As the chart shows, divorce rates vary significantly by state. The divorce rate in Massachusetts is half that of Delaware and about a third of that in Arkansas; Nevada, not surprisingly, is the divorce capital of the Nation.*

So what explains the wide variation in states? Cold weather states seem to keep people together a little better, though Hawaii does well (18th) and Alaska poorly (37th). Wealthier states seem to do somewhat better, though not universally so. The best indicator of a state's divorce rate appears to be political—though surprisingly, the left-leaning states have far lower divorce rates than the conservative, "family value" states. If you look again at the chart, you'll see states that were won by Al Gore in bold. These include nine of the ten states with the lowest divorce rates, and only one of the ten states with the highest divorce rates.

So, if you renounce that estate tax repeal, sell the SUV, and pick up a copy of *Earth in the Balance*, the odds are you'll start racking up the anniversaries!

THE ODDS OF CELEBRITY DIVORCE

Read the tabloids and you'll hear every week about celebrity divorces, with the emphasis on divorce recidivists—the Elizabeth Taylors, Kelsey Grammers, and Lisa Marie Presleys. But is that an accurate picture? Or is the silent majority more like Paul Newman and Joanne Woodward, happily yet quietly married for decades?

Well, because we thought you had a right to know, we conducted a comprehensive first-of-its-kind study of celebrity divorce. The study focused on Golden Globe award winners, who should fairly represent celebrities as a whole. (The Golden Globe categories include both tele-

* To be fair, the data are by state of occurrence rather than by state of residence, so if a Massachusetts resident gets a quickie divorce in Nevada (which certainly happens more than the reverse), the divorce counts against Nevada's record.

vision and motion pictures.) All 2001 and 2000 Golden Globe nominees and winners in the acting categories were included in the database, for a total of 151; due to multiple nominations and awards, the final yield was 127 unique celebrities.

Of the 127 celebrities listed, 68 percent had been married at least once. Billy Bob Thornton takes the honors of having the most marriages with five (as well as the most divorces, also with five—of course, by law it had to be at least four divorces, but you never know with Billy Bob).

Of the celebrities who had been married, the average number of marriages was 1.4, while the average number of divorces was 0.65. The probability of a celebrity marriage surviving is only 35 percent, so the odds are 1.9 to 1 against the marriage.

As bad as these numbers look for celebrity matrimony, they are almost certainly biased significantly downward. Most Golden Globe recipients are younger celebrities, some of whom have yet to ditch their loyal first wife or husband for the trophy wife or boy-toy husband befitting their newfound stature. (Jennifer Garner, a Golden Globe winner for her role in *Alias* in 2002, ditched her husband within a year. No word on a replacement.) Conversely, they have yet to suffer the inevitable descent into prescription painkillers (or worse) that tends to coincide with the cancellation of their hit show, which inevitably ends with a divorce and quick remarriage to a drug counselor, key grip, or other "stand by your celebrity" type. (Bear in mind: Jennifer Garner turns forty in nine years.) So, assuming that the odds thus far have been akin to what we saw at the tenth anniversary mark in the Census Bureau table, then the odds of a celebrity marriage surviving for a lifetime are probably 3 to 1.

Now aren't you glad nobody's ever heard of you?[2]

PART 2

CLIMBING THE LADDER OF SUCCESS

ACHIEVING SAINTHOOD

Your chances of becoming a saint have varied greatly over the ages, but here's the good news: we're now in a boom time for sainthood. Pope John Paul II exceeded all previous saint-making records, and set a standard against which all future popes will be measured. On His eighty-third birthday, in 2003, His Holiness canonized his four hundred seventy-third saint—a total almost equal to that of all popes over the previous *five hundred years.*

There have even been allegations that Pope John Paul II was running a "sainthood factory," hoping to popularize the Church in tough times (not unlike baseball owners boosting home run totals with a juiced ball and juiced ballplayers). In 1999, His Holiness beatified* (the step immediately before sainthood) Padre Pio, despite claims that he liked the intimate company of young women who wore perfume, and that the stigmata wounds that were his claim to fame were self-inflicted using acid. He also beatified Josemaria Escriva de Balaguer, the founder of the Opus Dei group (an arm of the Vatican dedicated to fighting heresy by all means necessary), who was a virulent anti-Semite and fascist sympathizer. Finally, in 2000, he beatified (after his predecessor had refused to) Pope Pius IX. In 1858, Pius IX ordered a six-year-old Jewish boy to be kidnaped from his parents and raised a

* Pronounced "bee-at-if-ied."

Catholic, on the grounds that the boy had been doused with Holy Water by a maid. He also had some interesting views on slavery, writing in his *Instruction 20* in June 1866, "Slavery itself, considered as such in its essential nature, is not at all contrary to the natural and divine law. . . . It is not contrary to the natural and divine law for a slave to be sold, bought, exchanged, or given." On the other hand, he endeared himself to future popes by formally decreeing the doctrine of papal infallibility.

The recent run on sainthood, though, is actually "back to the future" for the Catholic Church. Historians have established four eras of saintmaking in the history of the Church. The first lasted from the beginning of the Church until A.D. 313. These were the days of delegated and liberal saintmaking, with the Vatican adopting a hands-off approach. Any community was free to recognize a *hagioi*—that is, a saint or holy one. There was a significant downside for the canonized, however, as the way *hagioi* were identified was through their martyrdom. To borrow a phrase, the only good *hagioi* was a dead *hagioi*. (And since Christianity was then outlawed by the Romans, there were always plenty of dead *hagioi* to go around.)

Indeed, thanks to the Romans, the investiture process was not complicated—the local people simply collected the remnants of the *hagioi*'s torn body and deposited it at a holy site. (Many of these sites are present-day basilicas.)

This initial sainthood boom ended with the Edict of Milan in 313, when Emperor Constantine declared that Christianity was no longer outlawed. Without *hagioi* lying in torn remnants around them, communities had a much tougher time deciding who deserved sainthood and who did not.

Thus, from 313 to 993, your sainthood would have rested

in the hands of the local bishop, who was called upon to make the judgment calls necessitated by the Romans' exit. These were in some ways good times for those seeking sainthood, as popularity was more important than martyrdom. Go to church, tithe religiously, maybe perform a miracle, and the next thing you knew you'd been designated your own feast day on the local liturgical calendar. Still, concern began to grow that some bishops were easier than others in this process.

In 993, the Catholic Church adopted a formal canonization process, which began the third era of saintmaking, lasting until 1983. Pope John XV approved the first papal canonization (of Bishop Ulrich of Augsburg in 993), using a method that resembled a court proceeding. There were "petitioners," "procurators," and "promoters of the faith." That last position became more commonly known as "the devil's advocate"—not because he was in league with Satan (like most lawyers) but rather because his job was to be a stickler, making sure that a proffered saint was the genuine article. The devil's advocates must have been good, because the pace of saintmaking slowed appreciably.

So, in 1983, Pope John Paul II adopted a new process. The devil's advocate was abolished and canonization simplified to three steps. The first is an investigation by the local bishop of your writings and works, which is reviewed by a panel of theologians at the Vatican. If the panel and the cardinals of the Congregation for the Causes of Saints sign off, you are proclaimed "venerable."

At this point, however, you need to produce a miracle. (Literally.) But here's the tough part: *you have to produce the miracle after you're already dead*. (This is really bad news, for example, for David Copperfield, because none of those disappearing airplanes—or for that matter, the marriage to Claudia Schiffer—counts for squat. Ditto for the '69 Mets, and Pauly Shore's agent.) A post-death miracle proves that you are in heaven and able to intercede for mortal folks. Since the Catholic Church believes that it is God who truly makes saints, and the Church's job is merely to identify them, intercessions from Heaven look like a really good lead. The only exception to this miracle requirement is—you guessed it—martyrdom.

Once you cough up a miracle, or proof of your martyrdom is established, then you are beatified. At this point, you are basically a local-area saint, and can be honored by those in your hometown. But you're still on the sainthood farm team, and the call from the big leagues awaits. To take that third and final step, you need to produce another miracle or, in the case of martyrs, a first miracle. At that point, you're in like Flynn. (Actually, Errol Flynn was no saint, to put it mildly.)

The Odds

So, what are the odds? While the Vatican does not keep an official ledger, there are around five thousand active saints that have some information about them on Catholic websites. Divide this number by the number of people who have lived since the beginning of Christendom (approximately 100 billion),* and you get odds of 20 million to 1.

* If you doubt how relatively young our little Earth is, consider that 6 percent of the people who have ever lived are alive right now.

Improving the Odds

How can you improve your odds? The best route to becoming a saint is . . . becoming a pope. Pope John XXIII and Pope Pius IX were both beatified in 2000, and they will join a long list—just about every Pope Tom, Dick, and Gregory has made the list. Being a bishop won't hurt your chances either.

Before making sainthood your goal, however, you need to recognize that there are potential drawbacks. The great advantage of sainthood is of course being forever remembered and even prayed to in tough times. But some saints have it better than others. Everybody thinks of Saint Christopher, the patron saint of safe travel, or Saint Peter, who mans the pearly gates of Heaven, or jolly old Saint Nicholas. (Don't tell the kids, but Santa Claus achieved sainthood by buying three girls out of prostitution—sort of the Richard Gere of his time.) But you could end up a modern-day Saint, Bernard the patron saint of agricultural laborers. And while you might guess that nothing could be better than being Saint Valentine, there's a reason those valentines are red: there have actually been three Saint Valentines, all of whom paid dearly for their fame. The first Saint Valentine was a priest and physician who was beaten with clubs and beheaded by the Romans on February 14, 270. The second Saint Valentine (the Bishop of Interamna) was scourged and decapitated by the Romans as well. The third Saint Valentine was martyred in Africa; while details are sketchy, those betting on a decapitation hat trick probably have the odds in their favor.

Finally, you might notice how dopey championship athletes are fond of saying, "They can't ever take this away from us," though of course no one would ever want to take their championship away from them, and instead everyone will be reminding them about it at card shows for the rest of their lives. But with sainthood, they really *can* take it away from you. In 1969, there was a major saint housecleaning. Past saints were investigated, and those who didn't measure up, or were found never to have existed, were scratched from the list. So, it might pay to keep those miracles coming.

Secular Sainthood: Being Awarded the Congressional Medal of Honor

In the United States, there is a secular equivalent to sainthood: winning the Congressional Medal of Honor. When President Harry Truman presented Sergeant Paul Bolden with the Medal of Honor in 1945, Truman told him: "I'd rather have the Medal of Honor than be President of the United States." No one doubted him, for Medal of Honor recipients enjoy for a lifetime a universal and genuine reverence that Presidents and ex-Presidents can only envy.

The Medal of Honor is the only military award that is worn around the neck rather than pinned to the tunic. Recipients also have the option of wearing a six-sided blue rosette on their civilian attire. Regardless of relative rank, all soldiers salute a Medal of Honor recipient first, with the recipient returning the salute. The citations of every recipient are available at *www.army.mil/cmh-pg/moh1.htm*. (They are well worth a read on a rainy day—or Veterans Day.) The most recent recipients of the award were two members of the Delta Force in Somalia; their heroics were detailed in the movie *Black Hawk Down*. Until then, no Medals had been awarded for action since the Vietnam War.

Each branch of the military establishes its own standards

for the award, though they are similar. The Army regulations provide:

> The Medal of Honor is awarded by the President in the name of Congress to a person who, while a member of the Army, distinguishes himself or herself conspicuously by gallantry and intrepidity* at the risk of his life or her life above and beyond the call of duty . . . The deed performed must have been one of personal bravery or self-sacrifice so conspicuous as to clearly distinguish the individual above his comrades and must have involved risk of life. Incontestable proof of the performance of the service will be exacted and each recommendation for the award of this decoration will be considered on the standard of extraordinary merit.

The Odds

There have been 3,459 Medals of Honor awarded. Given that approximately 40 million Americans have served in the armed services since the Civil War, the odds of a soldier winning the Medal of Honor are about 11,000 to 1.

Improving the Odds

As with sainthood, a willingness to sacrifice one's life is a prerequisite to the award. About one in six Medals of Honor are awarded for conduct where the recipient was killed in action. Remarkably, at least two soldiers have lived to receive the award after throwing themselves on the proverbial grenade: one was fortunate enough to land on a

* If you love words, you've just got to love "intrepidity."

dud; the other had the presence of mind to throw his helmet on it first.

There are other parallels to sainthood. From the time the Medal of Honor was created in 1861 (for the Navy) and 1862 (for the Army) until World War I, the Medal of Honor was the *only* medal awarded to members of the armed services. As a result, standards for receiving the award varied greatly, not unlike the first era of saintmaking. As a result, a comprehensive review of recipients in 1917 led to 910 Medals being rescinded. New medals—for example, the Silver Star (originally the Citation Star), the Distinguished Service Cross, and the Navy Cross—were initiated for service that was extraordinary but short of that required to earn the Medal of Honor.

Only one woman has ever won the Medal of Honor. That woman was not Meg Ryan in *Courage Under Fire* or, as many people believe, Karen Emma Walden, the entirely fictional character she portrayed. It was Dr. Mary Walker, the first woman to serve as a doctor with the Army Medical Corps and the first American woman to be a prisoner of war. She won the Medal of Honor at the Battle of Bull Run in 1861 for performing her medical duties. Her medal was one of the 910 rescinded in 1917, but she refused to return the medal when the Army demanded it. Her Medal was restored by President Carter in 1977, after a campaign by her granddaughter.

Until relatively recently, no one would give you odds for winning the award if you were a member of a minority group. In 1991, President Bush honored for the first time a black veteran of World War I; in 1998, President Clinton presented the first Medals of Honor to black veterans of World War II. Asian-Americans who served gallantly in World War II had to wait half a century for recognition; President Clinton awarded the Medal of Honor to thirty-two of them in 2000.

In many ways, though, the Medal of Honor has been a remarkably egalitarian award. It is awarded regardless of rank. Indeed, eligibility for the award was originally limited only to enlisted men, with officers added later. Generals do not award it to themselves. Only one President—Theodore Roosevelt—has been awarded the Medal of Honor, and he seems to have earned it in his charge up San Juan Hill. Even then, he had to wait over a hundred years (until 2001) to receive it.

So, if you see an old man with a little six-sided blue rosette on his lapel, stop to thank him, and to ask him his story.

Getting into an Ivy League College

Intelligent, high-minded folks wearing sweaters and punching each other on the shoulder in a joshing sort of way. Summers at the Cape. Internships with one's friends' parents. Eating clubs and secret clubs and the quiet satisfaction of excluding others. These are the joys of the Ivy League. Not to mention, of course, the actual ivy, which conveys a certain history and tradition, though it eventually corrodes the underlying brick.

Ironically, given the Ivy League's academic reputation, its origins are football. The first reference to the "Ivy colleges" came in 1933 from a sportswriter, Stanley Woodard of the *New York Herald-Tribune*. Two years later, in 1935, AP sportswriter Alan Gould referred for the first time to the "Ivy League." Ironically, at that time, the phrase was a misnomer: the schools described by Woodard and Gould were at that point independent of any conference (which is why they needed a collective moniker). The phrase also included not just the current Ivies but also Army, and later Navy, which were also independents. The Ivy League was not formed officially—albeit without the service academies—until 1945.

The Odds

Many parents dream of their child attending an Ivy League school, yet the obstacles are large. High academic standards. Exorbitant tuition. And just that little bit of intimidation—the feeling that you're not quite up to it.

But here's some good news: Cornell is in the Ivy League!

The odds of being admitted to an Ivy League school are simple to calculate, as shown in the attached table:

Ivy League Admissions Statistics 2001–2002

School	*U.S.News* ranking	Applicants	Acceptances	Admit rate	Out-of-state	SAT 25–75%
Princeton	1	14,289	1,679	12%	85%	1380–1560
Harvard	2	19,014	2,110	11%	85%	1380–1570
Yale	2	14,809	2,038	14%	92%	1360–1540
Penn	4	19,153	4,132	22%	82%	1310–1490
Dartmouth	9	9,719	2,220	21%	98%	1330–1510
Columbia	10	14,094	1,729	12%	75%	1320–1510
Cornell	14	21,519	5,861	27%	55%	1270–1470
Brown	17	16,606	2,729	16%	94%	1290–1490

As noted, Cornell is definitely the best bet, because it has the highest admission rate, the largest class, and the lowest SAT scores. Cornell generally admits more students than Harvard, Yale, and Columbia *combined*. But note that only 55 percent of Cornell's incoming class comes from outside of New York, so unless you're from the Empire State, Cornell's ivy walls remain hard to scale. Next best bets are the University of Pennsylvania and Brown University, which combine relatively large classes and relatively forgiving admission rates.

These odds, though, are becoming increasingly deceptive, as are most statistics related to college admissions. The major reason is the much-loved (by parents) and much-loathed (by administrators) *U.S. News & World Report* rankings. Generally considered the definitive objective measure of the relative quality of schools, the rankings are widely reported and seriously affect a college or university's ability to recruit the best students. For that reason, of course, they are ruthlessly manipulated.

One primary factor in the rankings—25 percent of the school's overall score—is the school's "student selectivity"—that is, how good the incoming students are, and how picky the school is in selecting them. The most important component of student selectivity is SAT or ACT scores of incoming students (40 percent of this factor), but 15 percent is the school's admission rate, and 10 percent is its "yield"—that is, what percentage of students are admitted, and what percentage of admitted students decide to enroll. A lower admission rate and higher yield improve the school's ranking. So, for example, there have been reports of smaller, less distinguished schools manipulating these numbers by denying admission to their best applicants, figuring that they would not attend anyway; the school thereby decreases its admissions percentage and increases its yield, making it a better school in the eyes of *U.S. News*. (As a result of this problem, and the growing push by schools to force students to participate in early decision or early action programs in order to increase the schools' yield, *U.S. News* announced in July 2003 that it would end the use of yield as a ratings factor.)

The Ivies, though, have less room to manipulate their yield, since a large percentage of the admitted students matriculate, including the best ones. Thus, the most effective way for an Ivy League school to affect its ranking is to *generate the largest possible pool of unsuccessful applicants*. So just because you get a solicitation in the mail, or a nudge from your guidance counselor, urging you to apply to your dream

Ivy, don't assume that means anyone has decided you're likely to be admitted.

Improving the Odds

When most people think of getting an edge in admissions, they immediately think of affirmative action—efforts to find spots for minorities. Certainly, minority applicants get some breaks, and will continue to do so under the Supreme Court's recent decisions in the University of Michigan cases. But affirmative action for minorities is hardly the only way for someone to get an edge in admissions.

There is also, for example, affirmative action for well-heeled white people, which takes the form of legacy preferences—that is, preferences for children of alumni. At many schools, the preference here is every bit as strong as for traditional affirmative action. Such preferences are justified, of course, on the grounds that they maintain alumni support and, more importantly, alumni giving. Telling Thurston Howell III that Thurston Howell IV will not be admitted is hardly conducive to soliciting that big donation for the new Thurston Howell III Science Building. As affirmative action for minorities has come under fire, legacy programs have felt increased pressure and even been formally discontinued at some schools. But a letter from a parental alumnus or alumna will never be ignored.

And it gets worse. As colleges and universities seek to build their endowments, they seek out and reward wealthy philanthropists—even if the philanthropist in question went to another college, or no college at all. According to an exposé by the *Wall Street Journal*—entitled, appropriately enough, "Buying Your Way Into College,"[3]—how much the robber baron with the ne'er-do-well son needs to spend generally depends on the size of the school's endowment—say $20,000 for a small, third-tier college but at least $250,000 and probably more than $1 million for a top-ranked school. (Ironically, the most aggressive user of this "nouveau riche admissions policy" has been the self-proclaimed "Harvard of the South," Duke University, which was built with tobacco money but is now apparently open to all comers.) But this sort of behavior is to some extent frowned upon in the Ivy League as, well, just a little gauche. In the 1960s, Yale famously declined admission to the son of its largest donor, Paul Mellon.

One way for the nonwealthy to increase their odds is through athletics. The Ivy League does not allow any athletic scholarships, but that is not to say that the Ivies do not offer preferential admissions to athletes. They do. And the more obscure the sport, the more difficult to fill the roster spot. So you might think about signing Junior up for fencing lessons or getting him on the school curling team.

Overall, perhaps the best advice comes from Jacques Steinberg, a former admissions officer who wrote a book about his experiences called *The Gatekeepers*. As he puts it, college admissions offices are trying to build a *community*, and your best chance of being admitted to an Ivy League (or any other) school is to demonstrate how you would add to that community. He reports, for example, that the Ivies are generally short on tuba players. But the best example is geographical diversity. Colleges want their students to meet kids from all over the country, and if that means admitting an Alabaman whose SAT scores are not quite those of some New Yorker, then so be it.

There is, of course, one final more drastic approach to improving your odds of getting into the Ivy League: work hard in high school.

WINNING A RHODES SCHOLARSHIP

The Rhodes Scholarship is the most prestigious distinction a graduating college student can receive. The winners pursue a graduate degree at Oxford University, with all tuition and fees taken care of, as well as a stipend to cover transportation to and from England. The total value averages approximately $28,000 per year. But putting the Rhodes Scholarship in dollar terms is like assessing the worth of an Olympic medal based on the market value of its gold, silver, or bronze. Both are priceless.

The Rhodes Scholarship was established in 1904 by the estate of Cecil Rhodes, in order to bring outstanding overseas students to the University of Oxford. The Rhodes Trust describes Mr. Rhodes as a "colonial pioneer and statesman." Of course, "colonial pioneer" is a rather suspicious phrase, and of course it turns out that Mr. Rhodes was the founder of DeBeers, the diamond company, and made his money through the ruthless exploitation of the African people.[4] At one point, he ruled over one quarter of the continent, and he was essentially the founder of apartheid. A big thinker, Mr. Rhodes once declared his intention to "bring the whole uncivilized world under British rule." Not surprisingly, there are not a lot of monuments to him in Africa. Indeed, Zimbabwe (once named Rhodesia by the eponymous-thinking Mr. Rhodes) has publicly debated whether to disinter him and throw his body into the river. So let's just say the scholarship is a little ironic.

The criteria for selection are set down in Rhodes's will. These include "high academic achievement, integrity of character, a spirit of unselfishness, respect for others, potential for leadership, and physical vigor." The goal is to find those students who will "esteem the performance of public duties as their highest aim."

In keeping with this mandate, the Rhodes Scholarship has become the Mount Everest of academic achievement. The winners are not only first-rate academically, but well mannered, athletic, and all around super. While Bill Clinton is the only Rhodes Scholar to have become President of the United States, winning a Rhodes is generally a guarantee of the good life. If you could have only one entry on your résumé, "Rhodes scholar" is probably the one to pick.

The Odds

So, what are the odds? And can you improve them? The answers, surprisingly, are: "That depends" and "Quite a bit."

We'll focus first on the process in the United States. While Rhodes Scholars are selected from around the world, the United States gets by far the largest share—thirty-two fellowships per year. The U.S. application process has three steps.[5]

1. The candidate must be nominated by his or her college.
2. Each nominee is reviewed by one of the fifty state committees of selection.
3. The nominees selected from each state are then interviewed by one of eight district selection committees. Each district committee selects four Rhodes Scholars.

Considering that there are about 1.2 million students receiving bachelor's degrees from four-year colleges annually in the United States, that makes a randomly selected U.S. college student's odds of winning a Rhodes about 37,500 to 1. Better than the lottery, but still pretty daunting.

Improving the Odds

That said, your odds may be quite a bit better, depending on where you grow up and where you attend college. Under the Rhodes Scholarship's rules, you have some choice in the state—and thereby the district—from which to apply. You can apply either from your home state or a state where you have attended college for at least two years.

So think of the Rhodes Scholarship as the education equivalent of the United States Senate. If you want to get elected from New York, you'll need sharp political skills, a lot of experience, and a huge war chest. If you want to get elected from Wyoming or Montana, then the bar is usually quite a bit lower.

So, when application time comes, you'll want to avoid District II, which includes Delaware, New York, Pennsylvania, and West Virginia. The combined population of this district is 33.9 million, including a lot of eggheads and three of the nation's top twenty universities.[6] Equally bad is District I—Connecticut, Maine, Massachusetts, New Hampshire, New Jersey, Rhode Island, and Vermont—with a combined population of only 22.2 million but six of the nation's top twenty universities.

The grass is far greener in District VI, however, which includes Iowa, Kansas, Minnesota, Missouri, Nebraska, North Dakota, Oklahoma, and South Dakota. Their combined population is 22.6 million, and the region includes only one of the nation's top twenty universities, Washington University. You may also covet District IV—Alabama, Arkansas, Florida,

Georgia, Mississippi, and Tennessee—where the best schools are Emory (number 18) and Vanderbilt (number 20).

If you're truly desperate, then there is a further option: move to another country, preferably one in the British Commonwealth. Your best bets are Canada, which receives eleven scholarships per year, and Australia, which receives nine. While that's significantly fewer than the United States, bear in mind that the population of Canada is 31 million and Australia 19 million. Assuming that their college population is the same proportion of the population as in the United States, that knocks your odds down to about 12,500 to 1 in Canada and 9,400 to 1 in Australia. And the beer is better.

Writing a Best-selling Book

The first best-seller list was established in the United States in 1895, by Harry Thurston Peck, a reviewer at the literary magazine *The Bookman*. While *The Bookman* is lost to history, *Publishers Weekly* began its own list in 1912, and that list is still considered the most accurate in the industry, as it surveys the largest number of bookstores. For the general public, however, "*New York Times* best-seller" is the ultimate plaudit, and the *Times* list should be your ultimate goal.

The *New York Times* lists fifteen fiction, fifteen nonfiction, and five "advice/miscellaneous/how-to" best-sellers per week. (Just to torture agents, publicists, and writers, the *Times* actually releases to the trade, and publishes on its website, the top fifty in the first two categories and the top fifteen in the third; such information cannot be used for "*New York Times* best-seller" purposes, however.)

The Odds

With thirty-five best-sellers per week, that would mean 1,820 *New York Times* best-sellers per year, assuming 100 percent turnover in each list each week. In fact, though, numerous books stay on their respective lists for weeks or even months at a time. For example, during 2002, there were a total of only 108 books on the fiction best-seller list, or 15 percent of the 780 you would find with 100 percent turnover. So assuming the same holds true for the nonfiction and advice/miscellaneous/how-to lists, that means an average of around 250 distinct *New York Times* best-sellers per year. Each year, on the other hand, an average of 55,000 trade books are published in the United States. (Trade books are distinguished from, and actually outnumbered by, "professional books," which are manuals and similar work-oriented publications.) Thus, assuming you can get your book published, your odds of writing a *New York Times* best-seller are 220 to 1.

But the odds can get better. Next step down from "*New York Times* best-seller" is "national best-seller." In the trade, this means that you are on a best-seller list maintained by another organization that surveys booksellers nationally. *Publishers Weekly*'s list would certainly qualify, but so would the lists of the *Los Angeles Times* or *Dallas Morning News*. The lists of these regional newspapers tend to favor local authors. The lists qualify as national best-seller lists because they include a few bookstores nationally, but they place a disproportionate emphasis on sales at local or regional bookstores. This advantage is offset, though, by the fact that LA denizens tend to write screenplays rather than books, and Texans tend to favor gunplay over wordplay.

The best news for would-be authors of a "national best-seller," though, is the *USA Today* list, which includes 150 books each week, and does not differentiate by type of book. (There is some bitterness among advice/miscellaneous/how-to authors toward the *New York Times* list, as the number 6 best-selling miscellaneous/how-to book, which just misses the *Times* list, frequently has sold more books than the majority of the listed fiction and nonfiction "best-sellers." The stakes are high enough that the editor in chief of HarperCollins recently wrote the *Times* to ask that a Harper business book, *Good to Great,* be moved from the

A/M/HT list to the nonfiction list; he succeeded, and the book became a *Times* best-seller.)

Sum up the *USA Today*, local/national, and *Publishers Weekly* lists, and your odds of having a "national best-seller" are worlds better than having a *Times* best-seller. Assuming we're now talking about one thousand books per year, your odds of writing a "national best-seller" are down to 54 to 1.

But many a paperback labeled "The Best-selling Book from Today's #1 Fitness Guru!" was never a *New York Times* best-seller or even a national best-seller. How is that? Because they're *lying*. You see, there's really no law against labeling your book a best-seller, even if it was a best-seller only in your own home. It's what they call "puffery" in the advertising business and won't get you locked up. Of course, you have to live with yourself.

Improving the Odds

So how are you going to get there? Here are some basic steps.

Step 1: Write a book. Well, obviously, you're going to have to write a book (or pay someone to write one for you). This is not difficult, though. Wrestler Mick Foley wrote a best-seller. Kathie Lee Gifford is a best-selling author. Dennis Rodman wrote one of the top ten best-sellers of 1996, *Bad As I Wanna Be*. (Sample passage: "From the outside I had everything I could want. From the inside I had nothing but an empty soul and a gun in my lap.") Don't worry: writing the book is the easy part.

Step 2: Pick a good title. Without a good title, you are pretty much sunk. In fact, it would not be irrational to allocate your time as an author equally between writing the book and thinking of a title for it.

What makes for a good title? As the story goes, Bennett Cerf, the cofounder of Random House, was once asked what the perfect title for a book would be. His response was *Lincoln's Doctor's Dog*. That pretty much touches all the best-seller bases. Conversely, according to Michael Korda (who wrote a distinctly *non*-best-selling book on best-sellers, called *Making the List*), the famous editor Robert Gottlieb (who has edited everyone from Joseph Heller to Bill Clinton) was once asked what the *worst* possible title for a book would be. His response was *Canada, Friendly Giant to the North*.

How important is a title? Just think about the yacht you'd be sailing if you had first thought of the title ______ *for Dummies*—with the blank first filled in with *DOS* but eventually branching out to include *Golf, Dating, Mac OS-X,* and just about every possible pursuit that leaves us feeling inadequate. Or if, lacking creativity but long on shamelessness, you had missed the *Dummies* boat but decided to settle for *The Complete Idiot's Guide to* ______ and put out books on the same topics as the *Dummies* people.

A great example of the power of the title is a book on getting venture capital funding that was published in 2001—after all the dot-coms had folded, the IPO market had been eviscerated, and there was no venture capital funding available. Who would buy such a book? No one—unless you named it *A Good Hard Kick in the Ass.*

In the past few years, the best titles have been ones that suggest the author will be saying hateful things about Bill and/or Hillary Clinton. The strategy is solid: conservative foundations will begin buying in bulk, and Limbaugh and O'Reilly will have you on the airwaves pronto to spread the word. While the bile has been reduced somewhat during the

Bush years, if Senator Clinton's ambitions should ever extend to the Presidency, her candidacy might become the biggest boost to the publishing industry since Gutenberg invented the movable-type printing press in 1450. (Hint to ambitious right-wing authors: think *Bill Clinton's Link to the Holocaust* and *Hillary Clinton: Serial Killer Exposed*.)

Step 3: Find a publisher. With title and book in hand, you'll need to find a publisher and, ideally, an agent. Here, the process gets a little nebulous, as a human factor enters the picture: namely, a subjective assessment of you and your book by potential agents (that's not the human part) and editors.

Still, even if all the agents and editors should "pass" on your book (that's publishing talk for "Go away"), you need not give up hope. Self-published best-sellers are unusual but by no means nonexistent. The *Rich Dad/Poor Dad* series of best-selling personal investing books was originally self-published. So, too, was the mystic, New Age *Celestine Prophecy*. The road is not pretty: loading your car up with books, trying to persuade local bookstores to stock it, hoping that a branch of Borders or Barnes & Noble will notice it and buy a few copies, hoping that it sells well enough to attract the notice of the national media. You will need to quit your day job, but if you truly believe, and have a large enough trunk, knock yourself out.

Step 4: Have a previous best-seller to your credit. No factor is more important to your writing a best-seller than your having *already* written a best-seller. You might note the catch-22 here, as everybody has to have a *first* best-seller before they can have multiple best-sellers. But the point is a valid one. Whereas in the past, the authors on the best-seller list changed every year, the publishing industry's emphasis on mass marketing and chain store sales means sticking with a proven winner. Consider that the number one best-selling fiction author of 1994, John Grisham, also had the best-selling book in 1995—and 1996 and 1997 and 1998 and 1999. Danielle Steele had one of the top fifteen best-selling books of 1990; two of the best-selling books of 1991, two of the best-selling books of 1992; one of the best-selling books of 1993; three of the best-selling books of 1994; two of the best-selling books of 1995; two of the best-selling books of 1996; three of the best-selling books of 1997; three of the best-selling books of 1998; and two of the best-selling books of 1999. She didn't do so badly in the 1980s either.

So the lesson is: (1) Your chances of making the best-seller list as a newcomer are smaller than they have ever been before; (2) Once you've made it, you are more likely than ever before to repeat that performance.

Step 5: Platform. Here's the really hard part. As we saw with the Foley/Gifford/Rodman examples, the easiest way to sell a book—particularly a nonfiction book—is by being famous. There are two different versions of fame that can sell a book, though. The first is broad name and face recognition, where they plaster your big puss on the cover of the book and all the people who love you buy a copy. Think *Leadership,* the best-seller by Rudy Giuliani. The second is regular access to audiences, whether through a newspaper column, a regular feature on a TV show, a radio show, or a newsletter you publish. This type of fame does not expose you to as many people, but it exposes you regularly to a dedicated cadre of potential readers. Think Suze Orman—of NBC, CNBC, QVC, and radio—and best-selling author of the despicably titled best-seller *The Courage to Be Rich* (among others).

Step 6: Luck. Although all these steps are helpful, in the end best-seller status depends a lot on timing. Even the worst Osama Bin Laden biography was destined to sell books in September 2001; conversely, you wouldn't have wanted to bring *Radical E: From GE To Enron—Lessons on How to Rule the Web* to print in March 2002.

Going into Space

A generation ago, astronauts were the world's heroes, NASA its most respected institution, and the space program the U.S. government's most popular endeavor. Space flight not only represented adventure but was also a vital part of Cold War competition between the United States and Russia. It is easy to forget (or never to learn) how cosmonaut Yuri Gagarin's trip into space sent the Free World into a panic. Every young boy dreamed of becoming the next Alan Shepard or John Glenn or, eventually, Neil Armstrong, and taking up the challenge himself.

Today, with the most obvious space goals conquered, astronauts generally are ignored by the broader population, except in the event of tragedy. Their collective identity has moved from daredevil pilot to hardworking scientist. Asked to name an astronaut, the first name that comes to mind for many is Christa McAuliffe, the schoolteacher who tragically perished with the Challenger.* The faces, if not the names, of the Columbia crew are now the new image of NASA.

But that does not mean that astronaut is not still among the most respected professions. When there is an astronaut in the room, people tend to migrate toward him (and, more recently, her). The magic of having seen the earth from above still hangs about this small and special group of people.

* Anyone considering a school outing or kid's birthday party should look up the Challenger Center, which is a living memorial that allows children to experience the excitement of space flight and learn the wonder of science; there are sites across the country. Check it out at *www.challenger.org*.

The Odds

So what are your odds of becoming an astronaut? The first steps are quite like applying to college. You fill out application materials and hope to be part of that year's freshman class. With NASA, classes generally start every two years, with school lasting two years. One difference from college: you have to declare your major before you enter school. Here, your choices are pilot (PLT), and eventually qualified career commander (CDR), or mission specialist (MS).

So what are your odds of making it into an astronaut training program? You probably want to be American or Russian, as those countries are the only ones that have regular training programs. And there are a few characteristics, some unfortunately immutable, that you will have to demonstrate. For NASA's program, for example, these requirements include:

- U.S. citizen
- bachelor's degree in engineering, physical science, mathematics, or biological science
- 20/200 vision, correctable to 20/20, for mission specialists; 20/70 vision, correctable to 20/20, for pilots
- 4' 10.5" to 6' 4" tall for mission specialists, and 5' 4" to 6' 4" tall for pilots (It is not clear whether NASA disdains short pilots for safety or aesthetic reasons.)
- if applying to be a mission specialist, three years of professional experience, though a master's degree counts for one year, and a doctorate counts for all three
- if applying to be a pilot, at least one thousand hours flying in command of a jet aircraft

In other words, if you're majoring in art history at Amherst, and find yourself squinting at the paintings from across the room, you can forget about it. If you're ROTC at MIT, then (in addition to being very lonely) you're astronaut material.

Beware: the application itself is daunting. Your average Joe or Jane is probably going to want to go the mission specialist route, since the pilot route is for a unique breed. But even the application for mission specialist is tough. NASA still would like to know whether you have combat flight experience, as well as your overall flight experience in hours. NASA feels compelled to add: "Include any experience, *except passenger* (e.g., pilot, copilot, crew member, test subject, etc.)." Presumably some applicants were treating their frequent flier miles as flight experience. ("Hey, you may have flown two tours off a carrier in the Gulf, but I've flown US Airways into Rochester, New York—*coach*.") That said, there may still be a loophole: given that commercial fliers are now searched, deprived of food, and (near Washington, D.C.) forbidden to get up to go to the toilet, they may fall within "test subject."

But if you did put in all the required hours at the science lab and can fit inside the cabin, what are your odds of being selected? Not great, but gradually improving, as the space shuttle program has had—and likely will continue to have, even post-Columbia—more room for astronauts, and flies more frequently than the old Gemini and Mercury craft, thereby requiring larger classes of astronauts.* Your odds are 33 to 1 (118 of 4,000) of being selected to participate in a week-long interview and medical evaluation. While you're not in the program at this point, you've at least gotten a free

* If you're planning to go the Russian route, though, beware that the classes often have themes. Thus, the 1962 Soviet class was the "female group" (subsequently, seen in *Playboy*'s special feature "Soviet Space Babes: You're Going to Love Reentry!"); the 1965 Soviet class included a journalist group. For both Russia and the United States, an entire class sometimes will consist of air force officers if a classified military mission is planned.

checkup and can dine out on your experience with friends and relatives for years. At this point, though, your odds are still about 5 to 1 against being selected for the program, as the entering class of late has been twenty people.

Once you graduate, you are in NASA parlance an "unflown astronaut." You are also considered, however, an "active astronaut," whom NASA defines as (1) an unflown astronaut in advanced training, or (2) a CDR, PLT, or MS who has been assigned to a flight or is eligible for flight assignment. Achieving the "active astronaut" designation is crucial for male trainees, *because this is when you become a chick magnet.* As of 2003, NASA counted 115 active astronauts: 80 were qualified CDRs, PLTs, and MSs, and 35 were unflown astronauts in advanced training. (The Russians had only 32 cosmonauts.)

But wait, there's a loophole—there's always a loophole. If you're not up for the rigor and vagaries of the NASA program, there's one major alternate route. Welcome to the world of "payload specialist astronauts." NASA describes payload specialists, somewhat cryptically, as "individuals selected and trained by commercial or research organizations for flights of a specific payload on a space flight mission." In other words, if a contractor sells NASA some fancy gadget, it gets the right to send up one of its own. NASA lists no height or vision requirements for these folks, of whom forty-two have entered space.

The Russians have lately opened up another avenue toward space: space tourist. The first to go up was billionaire Dennis Tito, who paid the Russians $20 million for a ride to the International Space Station in April 2001. Then, teen heartthrob Lance Bass of *NSYNC flirted with the idea before parting ways with the Russians. NASA's own experiment with space tourists began and likely ended with Christa

McAuliffe; whom NASA records now sadly list, uniquely, as a "space flight participant."

In any event, if you're bound and determined to get into space, don't plan on making a billion dollars and heading out to Russia. Study hard and go to Cal Tech.

Winning an Academy Award

Dreaming of winning an Oscar but realize that you have neither the face nor the personality to become an actor? Here's some good news! While the Oscars include twenty-three award-winning categories, only four are for actors. Moreover, each of the acting categories has only one winner,* whereas if a half dozen nerds end up sharing the award for Best Sound, the Academy simply hands out six trophies. So actors actually end up with fewer than 10 percent of the awards. That doesn't mean that you have a better shot than Gwyneth Paltrow, but you needn't give up hope, either.

The Odds

So what are the odds? The number of Oscars awarded varies each year, since, as noted, many categories can have multiple winners. For the last ten years the number of Oscars awarded has varied from thirty-three to forty-one for the twenty-five official categories, with an average of thirty-eight awards. (There are other awards such as the science and technology

* While there have been two ties in the history of the Oscars, this frequently is not sufficient to affect the odds materially. (If you must know, Wallace Beery [*The Champ*] and Frederic March [*Dr. Jekyll and Mr. Hyde*] tied for Best Actor in 1931, and Katharine Hepburn [*The Lion in Winter*] and Barbra Streisand [*Funny Girl*] tied for Best Actress in 1968.)

awards, honorary awards, and special achievement awards, but their winners do not always receive an actual Oscar. So the heck with that.)

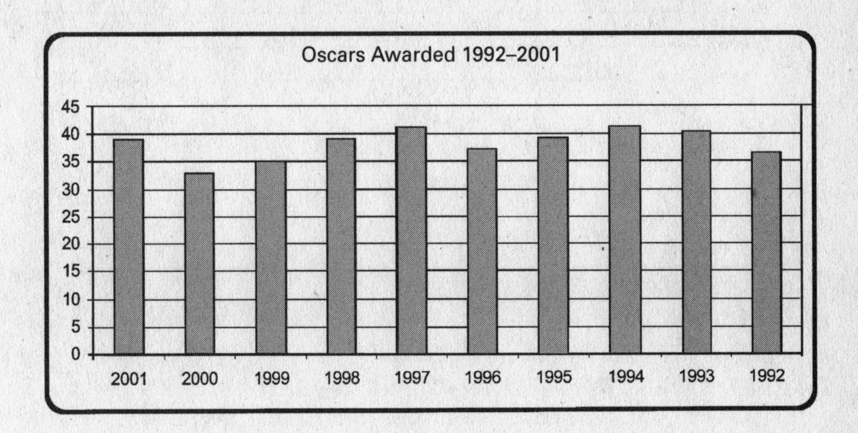

So, who is eligible for these thirty-eight-odd Oscars? They don't give Oscars to plumbers, farmers, or lawyers. According to the Bureau of Labor Statistics, there are approximately 440,000 people working in the trades that qualify for Oscars. So, enter show business, and generally speaking, the odds of receiving one of those thirty-eight-odd Oscars next year would be around 11,500 to 1.

Improving the Odds

But we can improve those odds a bit with some wise career choices. Below is a chart showing the size of each professional group vying for that Academy Award.

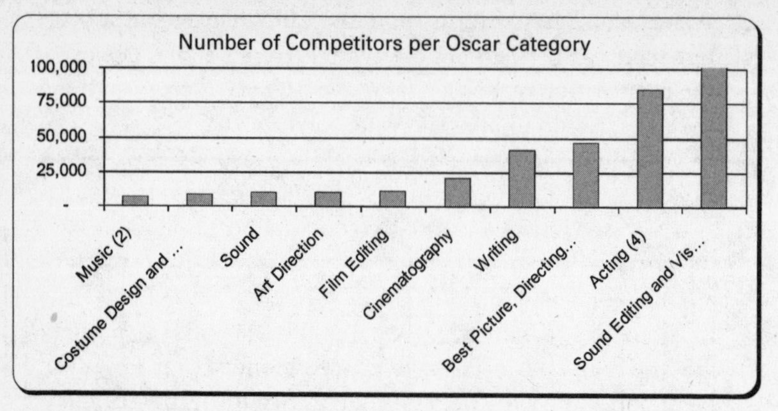

Source: Bureau of Labor Statistics

Surprisingly, the best odds of winning an Oscar belong to music directors and composers (not exactly the first jobs that come to mind when you think of the movie business). The guild is not very large, and there are often multiple Oscars awarded each year. If you are a music professional and are involved in a movie project, your odds of winning an Oscar are 1,666 to 1 each year.

In contrast, the odds for some dopey actor are 20,000 to 1. But acting isn't the toughest way to win an Oscar. The fiercest competition is among the almost 200,000 movie professionals vying for about six Oscars per year for sound editing and visual effects. For these poor wretches, the odds are about 33,000 to 1. Plus they have to work in dark offices and never get free clothes!

Of course, these are your odds each year, so over the course of, say, a forty-year career, you odds are much higher—roughly 40 to 1 for the musically inclined, and 500 to 1 for the thespians. That assumes, though, that you don't run up against an Oscar hog like Walt Disney (twenty-six Oscars, including twelve for best short subject) or Edith Head (eight Oscars for costume design) or John Williams (five Oscars for Best Score).

So once you've chosen a profession, is there anything more you can do to improve your odds?

Well, in recent years it has been increasingly possible to buy an Oscar. It all began in 1999, when the odds-on favorite to win Best Picture was the Steven Speilberg classic *Saving Private Ryan*. Among the other nominees was a fun but forgettable little movie called *Shakespeare in Love*. But Miramax, the production company behind *Shakespeare in Love*, came up with a novel strategy: they pretended that the Oscar was a political campaign, and they were a tobacco PAC. They spent lavishly—the true number is secret but rumors varied from $6 to $15 million—on advertising. They allegedly even hired retired Academy members to phone bank voters. Asked to respond in kind, Spielberg announced that he would take the high road—and promptly suffered the same fate as every politician who has ever taken the high road. Now film students watch *Saving Private Ryan*, but the little statue sits at Miramax.

But suppose you don't want to have to rely on devious studio bosses to promote your deserving efforts. How about a grassroots effort, emphasizing personal contacts and word of mouth? In 1935 cinematographer Hal Mohr, denied a nomination for his work on *A Midsummer Night's Dream*, launched a write-in campaign—and won. To honor Hal's historic victory, the Academy promptly prohibited write-in votes. Why? Well, because they're evil, that's why!

A Final Thought

If all this advice pays off and you do manage to win an Academy Award, you will need to make an acceptance speech. So, as a final Oscar-related service, here are two key themes to emphasize if you want your speech to advance your personal and professional life and avoid causing you ridicule: (1) humility, and (2) brevity. If you want to raise the odds of public humiliation, then you should talk about: (1) yourself, or (2) politics. To illustrate the power of these themes, here are the best and worst Oscar acceptance speeches ever given:

Worst (a tie)

- Cher, Best Actress for *Moonstruck,* 1987: "I'd like to thank everyone I worked with on the movie. They were really fabulous. It was really a great experience for me. My makeup man, who had a lot to work with. My hairdresser. My assistant . . ."
- Vanessa Redgrave, Best Supporting Actress for *Julia,* 1977: Redgrave praised Hollywood for having "refused to be intimidated by the threats of a small bunch of Zionist hoodlums . . ."
- Sally Field, Best Actress for *Places in the Heart,* 1984, "I wanted more than anything to have your respect. The first time [when she won the Oscar in 1979 for *Norma Rae*] I didn't feel it, but this time I feel it and I can't deny the fact that you like me. Right now, you *like* me!"

Best (also a tie)

- Jessica Yu, Best Short Subject Documentary, 1996: "What a thrill. You know you've entered new territory when you realize that your outfit cost more than your film."
- Maurice Jarre, accepting Best Score for *A Passage to India* in 1985—the year *Amadeus* won Best Picture: "I was lucky Mozart was not eligible this year."

Remember, too, how golden silence can be: in 2003, Michael Moore received a chorus of boos from an indignant academy when he used his acceptance speech for *Bowling for Columbine* to crusade against the Iraq war; Roman Polanski, barred from entering the country for drugging and raping a fourteen-year-old girl, and thus saying nothing, received a thunderous standing ovation for the *The Pianist*.

Growing Up to Be President

Estimating your chances of becoming President depends largely on whether we feel future Presidents will look like past Presidents. Based on past experience, one might assume that women and racial and ethnic minorities have zero chance of becoming President, but the odds are certainly better than that going forward. We'll take a look at which Presidential trends have proven the most durable, and seem likely to repeat in the future.

The Odds

The odds are self-evident here: the United States has over 280 million citizens and only one President at any given time. Given that your prime Presidential years are roughly ages forty to seventy-two, you have only eight potential administrations to lead. So, figure your odds at around 10 million to 1. Those are long odds indeed. But is there a group for whom the odds are shorter?

Improving the Odds

Well, first, keep in mind that the easiest way to become President is to get elected (or selected) Vice President. Ten sitting Vice Presidents—John Adams, Chester Arthur, Teddy Roosevelt, Millard Fillmore, Gerald Ford, Andrew Johnson,

John Tyler, Harry Truman, Lyndon Johnson, Richard Nixon, and George H. W. Bush—have moved up to the big leagues. Ford wasn't even elected Vice President; he was selected by crooked Richard Nixon to replace crooked Spiro Agnew. Andrew Johnson was by all accounts an awful person, and Arthur was a mediocre politician (though a pretty good President).

While most folks are aware of the ability of Vice Presidents to move up, the odds are almost as good if you are Secretary of State. Six Secretaries of State—Jefferson, Madison, Monroe, Quincy Adams, Van Buren, Buchanan—have become President. While none have moved up recently, Colin Powell would not be a bad bet someday. (His immediate predecessor, Madeline Albright, is foreign born and therefore ineligible.)

There is no doubt, though, about the best place from which to launch a Presidential bid. Seventeen governors from nine states have become President. Nor is there doubt about the worst. Historically, Presidential bids launched from the Senate have proved disastrous. Only two men, Warren Harding and John Kennedy, have moved directly from the Senate, even though it attracts quality people and gives them a national stage.

Height also plays a major factor. Eighteen Presidents have been six feet tall or higher, an extraordinary fact given that average heights were significantly lower in the nineteenth and eighteenth centuries. Height also seems to correspond with success in office. Lincoln was the tallest President, at 6' 4½", while Thomas Jefferson was a remarkable 6' 2½" when he became President in 1801—the Manute Bol of his time.

In a country as religious as the United States, you would expect the most popular religions to produce the most Presidents. In fact, there has been remarkable variety in religious preference.

Agnostic/Atheist/Deist/ Unaffiliated
- Thomas Jefferson
- Andrew Johnson
- Abraham Lincoln

Baptist
- Jimmy Carter
- William Clinton
- Warren Harding
- Harry Truman

Congregationalist
- Calvin Coolidge

Disciples of Christ
- James Garfield
- Lyndon Johnson
- Ronald Reagan

Dutch Reformed
- Theodore Roosevelt
- Martin Van Buren

Episcopalian
- Chester Arthur
- George H. W. Bush
- Gerald Ford
- William Henry Harrison
- James Madison
- James Monroe
- Franklin Pierce
- Franklin Roosevelt
- Zachary Taylor
- John Tyler
- George Washington

Methodist
- George W. Bush
- Ulysses Grant
- Rutherford Hayes
- William McKinley

Presbyterian
- James Buchanan
- Grover Cleveland
- Dwight Eisenhower
- Benjamin Harrison
- Andrew Jackson
- James Polk
- Woodrow Wilson

Quaker
- Herbert Hoover
- Richard Nixon

Roman Catholic
- John Kennedy

Unitarian
- John Adams
- John Quincy Adams
- Millard Fillmore
- William Taft

Episcopalians seem to have the best shot, but the differences don't seem worth a religious conversion. Still, if you aren't praying to some God, it's probably a bad sign that no atheist or agnostic has been elected for over 140 years.

When it comes to secular education, you're probably going to want to go back and thumb through that chapter on getting into the Ivy League. Harvard College has produced six Presidents: Adams, Quincy Adams, Roosevelt (Teddy), Roosevelt (Franklin Delano), Hayes, and Kennedy. Yale College has produced Taft and the two Bushes. Your next best bet is probably a service academy, as West Point produced Grant and Eisenhower and the Naval Academy, Carter.

You may also wish to give careful attention to your profession. Three in particular will offer you a big Presidential leg up—two of them are relatively predictable, the third not.

- The biggest advantage goes to that group of people we love to hate: lawyers. Twenty lawyers—from Adams to Clinton—have gone on to become President.
- Professional soldiers have fared well in the Presidential sweepstakes, with Jackson, Harrison (William Henry), Taylor, McKinley, Roosevelt (Theodore), and Eisenhower all making the move to commander-in-chief. Others, while not professional soldiers, nonetheless had what the British call "a good war": James Monroe, Franklin Pierce (who rose from the rank of private to brigadier general in the Mexican War), John F. Kennedy, and George H. W. Bush.
- Schoolteachers, oddly enough, have also fared extremely well. John Adams, James Garfield, Chester Arthur, and Lyndon Johnson—all these Presidents began their careers addressing small, very young audiences. While not a schoolteacher, Woodrow Wilson was a professor and president of Princeton University. And Millard Fillmore married his schoolteacher.

Oddly, once you choose a profession, you are best advised to abandon it as quickly as possible, and begin pursuing politics. The list of businessmen Presidents is remarkably short. Most modern Presidents never worked in business.

Most practiced law (Nixon, for a long while, Clinton for a short while) or moved immediately into politics (Kennedy, Johnson). Carter was a successful businessman, and George W. Bush an unsuccessful one; both stressed their gubernatorial rather than business careers when they made the big run. It is odd, given America's capitalist economy, but the only person to run for President on the strength of a business career in recent memory was Ross Perot, and that did not turn out well at all. Anyone remember the "Draft Lee Iacocca" movement?

Summing Up. So, if you're bound and determined to grow up to be President, here's your flight plan:

1. Attend Harvard, or at least Yale. You needn't excel. Student government would probably help. You can drink all you want. These are "youthful indiscretion" years.
2. After graduation, join the Teach for America program and (without having to take any education courses) teach some underprivileged kids for two years. *Remember to have someone take pictures of you teaching the underprivileged kids.*
3. Then, enlist in the military. Branch doesn't matter. You can even go AWOL occasionally. But make a few buddies who'll appear in your campaign video. If there's a war on, you may wish to do something heroic, but this is strictly optional.
4. Somewhere along here, get married. You needn't be too picky—hey, you're not going to find anybody crazier than Mary Todd Lincoln—but someone who can make a stump speech wouldn't hurt. Try to propose in a way that is easily and humorously retold.
5. Go to law school. Again, you needn't excel, but you need to work seriously at demonstrating leadership and charm to your classmates, as they are your future fundraisers and assistant secretaries.

6. Begin attending church. If you're within shouting distance of Episcopalian (say, if you're Catholic or Presbyterian), you might want to convert, but don't make too big a leap, or you'll look insincere. Get comfortable saying the words "and God bless the United States of America."
7. Get into politics as fast as possible, and set your sights on becoming governor.

May the odds be with you.

PART
3

REALLY BAD STUFF

Dying by Accident

When we think about beating the odds, we often think of avoiding the fatal accidents that we know happen every day but try to ignore. The National Safety Council thoughtfully tracks the odds of dying of a variety of non-natural causes—everything from choking on your own vomit (thank you, Jimi; thank you, Elvis) to "excessive heat or cold of man-made origin."[7] We'll look at the area with the shortest odds—transportation deaths—in a later section and here focus on the odds of the remaining accidental deaths.

The Odds

Here are the odds of accidental death. The categories used by the National Safety Council are based on those established in the World Health Organization's International Classification of Disease (though most have nothing to do with disease).

The Odds of Dying By Accident

Type of accident	One-year odds	Lifetime odds
Falls	20,728 to 1	270 to 1
Exposure to inanimate mechanical forces	99,606 to 1	1,299 to 1
Exposure to animate mechanical forces	1,274,860 to 1	16,621 to 1
Accidental drowning and submersion	77,308 to 1	1,008 to 1
Other accidental threats to breathing	49,577 to 1	646 to 1
Exposure to electric current, radiation, temperature, and pressure	569,562 to 1	7,426 to 1
Exposure to smoke, fire, and flames	81,487 to 1	1,062 to 1
Contact with heat and hot substances	2,218,049 to 1	29,918 to 1
Contact with venomous animals and plants	4,472,459 to 1	58,311 to 1
Exposure to forces of nature	183,347 to 1	2,390 to1
Accidental poisoning by and exposure to noxious substances	22,388 to 1	292 to 1
Overexertion, travel, and privation	1,428,377 to 1	18,623 to 1

Source: National Safety Council

Now, as you might expect given that they were negotiated at an international organization, these categories are a little tough to parse. So, we'll break them down a bit.

Falls (270 to 1). There are lots of different ways to fall to your death. Of those subcategorized by the National Safety Council, the most likely is a fall down stairs, with lifetime odds of 2,503 to 1. Surprisingly, the odds of falling out of your bed, chair, or "other furniture" are also relatively short: 5,700 to 1.

Exposure to inanimate mechanical forces (1,299 to 1). Here, the most likely killer is the accidental discharge of a firearm (4,317 to 1), which basically means "dumb guy cleaning gun" or "drunk hunter." The somewhat oblique "struck by or striking against object" presents odds of 4,224 to 1, and presumably incorporates asteroids and lawn darts. While the odds of your dying from a "foreign body entering through skin or natural orifice" are a reputable 93,605 to 1, we're not even going to think about what the second half of that means. Most of the inanimate mechanical forces seem to relate to the hazards of manufacturing: for example, "being caught between objects" (a shudder-producing 38,247 to 1); "explosion and rupture of pressurized devices" (107,787 to 1); and the innocuous-sounding but deadly "contact with machinery" (5,719 to 1).

Exposure to animate mechanical forces (16,621 to 1). Here you have the risk of being bitten to death by a dog (a pretty low 142,279 to 1, given the amount of worrying we do about it). Your odds of being fatally struck or bitten by another mammal is 51,550 to 1, so look out for those raccoons and rats, not to mention rampaging hippopotami.

Accidental drowning and submersion (1,008 to 1). Here all our (and Natalie Woods's) concerns appear to be well justified. The water is not our friend. So your odds of drowning in a swimming pool are a very real 6,711 to 1. The odds are even worse for ponds, lakes, and oceans, coming in at a collective 2,935 to 1.

Other accidental threats to breathing (646 to 1). This World Health Organization category is particularly opaque, as no longer breathing seems more like a result than a cause. So what's up here? Mostly, choking to death, where the odds of 5,558 to 1 should have every steakhouse training the waiters in the Heimlich maneuver. The odds of choking on nonfood objects are even greater (1,258 to 1), and then for rock stars there's your own gastric contents (vomit) at 8,530 to 1.

Exposure to electric current, radiation, temperature, and pressure (7,426 to 1). Here, as you might guess, we're mostly talking about electrocution in all its various forms, with power lines (28,008 to 1) the most likely culprit.

Exposure to smoke, fire, and flames (1,062 to 1). The primary risk is a house or building fire (1,329 to 1). There is a surprisingly high risk of ignition or melting of your clothes during daytime (31,759 to 1); nightwear is significantly safer (592,829 to 1).

Contact with heat and hot substances (28,918 to 1). Distinguished from, and less threatening than, smoke, fire, and flames, your heat and hot substances presumably include molten lava and McDonald's coffee.

Contact with venomous animals and plants (58,311 to 1). Decades ago, one-hit wonder Jim Stafford sang, "I don't like spiders and snakes," but if he'd known the odds, he would have focused his attention on hornets, wasps, and bees. The odds posed by those buzzing creatures (82,720 to 1) dwarf those posed by venomous snakes and lizards (508,139 to 1) and venomous spiders (592,829 to 1).

Exposure to forces of nature (2,390 to 1). There is symmetry here, as the danger of heatstroke (5,988 to 1) is roughly the same as the danger of bitter cold (5,948 to 1).

Accidental poisoning by and exposure to noxious substances (292 to 1). Here the numbers are grim, but behavior driven. Drug overdoses come in at 592 to 1; alcohol poisoning (which doesn't include all the other adverse effects of alcohol) is 11,116 to 1. Surprisingly, "nonopioid analgesics, antipyretics, and antirheumatics" pose a substantial risk, at 21,172 to 1—"surprisingly" because who knew those were even real words?

Overexertion, travel, and privation (18,623 to 1). No details are given for this category. The three items don't even seem to have much in common, except for a certain family trip to Venice in 1994 with various in-laws in tow.

Summing up. Sum them all up, and your odds of dying in a non-transportation accident—as opposed to disease or other natural causes—are about 69 to 1.

Improving the Odds

Contributing Factor—Sex

Two of the prime determinants of accident-proneness are age and sex. Men are twice as likely to die from accidental deaths at all ages up to sixty-five.

Estimated Deaths per One Thousand Persons Over Next Ten Years

Age	Accidental death rate	
	Women	Men
20	2	5
25	1	5
30	2	4
35	2	5
40	2	5
45	2	5
50	2	4
55	2	4
60	2	5
65	3	6
70	5	7
75	7	11
80	14	19
85	15	21

Source: Dr. Steven Woloshin, Dr. Lisa M. Schwartz, Dr. H. Gibert Welch, "Risk Charts: Putting Cancer in Context," *Journal of the National Cancer Institute,* June 5, 2002.

The reason for the higher accident rate is simple: men are stupid. Gradually, they get less stupid, but never become as smart as women. As for age, accident rates for each sex are fairly constant until retirement age, at which point they steadily increase.

In sum, if you're a woman, the odds of accidental death in your twenties, thirties, forties, and fifties are 500 to 1.

For men, the odds are 200 to 1 until you reach your fifties, when they drop to 250 to 1 for a blissful ten years. For both men and women, your sixties and older are a time to start padding all surfaces and letting someone else do the driving.

Transportation Deaths

The odds of all the wacky accidents described above are dwarfed by one category of accident: transportation. Here, the odds are only 77 to 1, or about the same as the odds of all other accidents *combined*.

Clearly, transportation is the area to focus on if you wish to reduce your odds of accidental death. (You have only limited control over whether you'll be bitten by a dog, and you don't even know what nonopioid analgesics are.)

Here, though, the National Safety Council data are not really of use, for while they give you the lifetime odds of being killed as, for example, a pedestrian (588 to 1) or a streetcar rider (3,556,975 to 1), that information is no use in deciding which means of transportation is safest. Walking is not 6,000 times safer than streetcar riding; it's just that a lot more people walk than ride streetcars. To decide how to travel, we need to understand the death rate *per unit of distance traveled*.

Thankfully, the British, who have very precise accident reporting regulations, keep close track of these matters. Here is what the data show over the past decade.

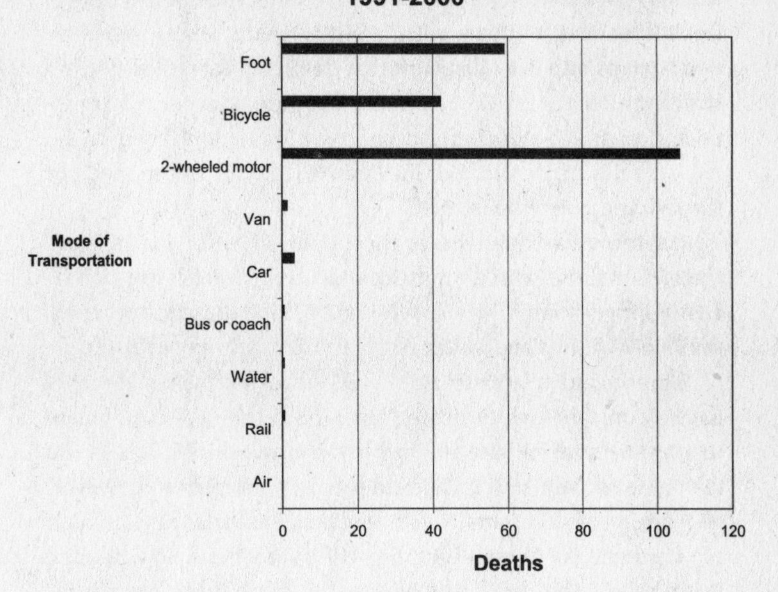

Since some of the death rates are barely a blip on the old graph, here are the hard numbers:

Deaths per Billion Passenger Kilometers 1991–2000

Method of Transportation	Death Rate
Air	0.02
Rail	0.49
Water	0.4
Bus or coach	0.4
Car	3.1
Van	1.3
2-wheeled motor	106
Bicycle	42
Foot	59

So what do we learn? While cars are certainly dangerous, *walking* is almost twenty times as dangerous. Bicycling is fourteen times as dangerous. (While that's largely because pedestrians and bicyclists are run down by cars, that doesn't really matter, does it?) The most dangerous means of transportation is "2-wheeled motor"—which would tend to be motorcycles in the United States and scooters and mopeds in the UK and everywhere else.

There are two qualifiers to these data. The first relates to car travel. For those who do not drive drunk, the odds are significantly longer than for those who do, making automobiles competitive with bus and water travel in the safety department.

The second relates to air travel. All other means of travel have a constant risk of death throughout the trip: e.g., you're just as likely to be run down by the proverbial truck in the fifth mile of a ten-mile walk as in the first. For planes, however, 90 percent of accidents occur on takeoff and landing, which are a requisite of every flight, regardless of length. So while the British show the death rate of flight to be roughly 2 per hundred billion kilometers, it is actually higher for shorter trips and lower for longer flights. Thus, using similar data, Professor Arnold Barnett of the MIT Sloan School of Management has calculated that the break point for choosing between driving and flying is 144 miles; so, for any trip under 144 miles, you're better off in a car. (This is particularly so if you are a rock musician or political candidate, who both seem to fare badly in these circumstances. Consider: given all the drinking and carousing that they do, when was the last time you heard of a musician or politician dying in an *automobile* crash?)

SURVIVING A TRAIN CRASH

This sidebar will change, and perhaps save, your life, for its message is very simple. Of all the accidents described in the previous chapter, there is one where the odds are in your hands—or, more aptly, in your seat. The key to surviving a train crash depends on one fundamental insight: *there are no rear-end train collisions.* Period. Trains crash when they run head-on into another train, a car, or a cow, or when a problem with the track causes a derailment. In every case, though, the cars most likely to suffer from impact, or to derail, are the cars *in the front of the train.*

Consider, for example, how events played out when an Amtrak passenger train hit a truck at a railroad crossing in Bourbonnai, Illinois, in 1999. The results, as detailed in the ensuing National Transportation Safety Board report, were unusually tragic—11 dead and 122 injured—but the physics were the same as always:

> The impact caused both locomotives and 11 of the 14 cars in the consist to derail. The lead locomotive came to rest at a point about 560 feet south of the grade crossing. Adjacent to this was the second locomotive, the baggage car, a transition sleeper car, a sleeping car, and the diner car, all of which were involved in a general pileup of wreckage in that area. (See figure 9.) The remaining cars in the consist either derailed and remained upright or did not derail at all. None was involved in the general pileup of wreckage. These cars did not display any evidence of serious carbody breach, and damage was mostly limited to the carbody end-structures.

The key parts are: "The remaining cars . . . either derailed and remained upright or did not derail at all . . . These cars did not display any evidence of serious carbody breach." That means that while passengers in the front of the train were dying, there were passengers in the back who walked off the train with their luggage—a not atypical outcome. Here's an NTSB diagram to make the point crystal clear:

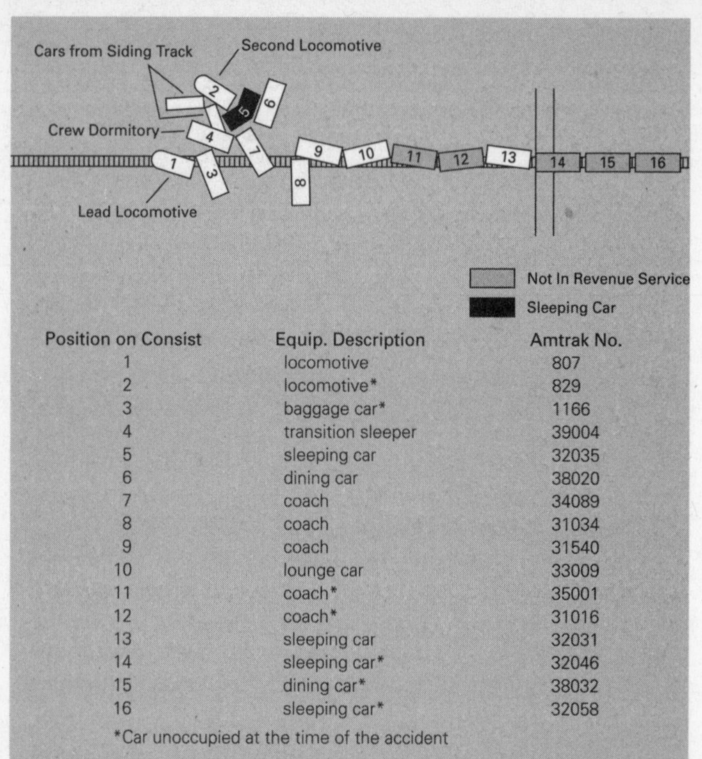

Position on Consist	Equip. Description	Amtrak No.
1	locomotive	807
2	locomotive*	829
3	baggage car*	1166
4	transition sleeper	39004
5	sleeping car	32035
6	dining car	38020
7	coach	34089
8	coach	31034
9	coach	31540
10	lounge car	33009
11	coach*	35001
12	coach*	31016
13	sleeping car	32031
14	sleeping car*	32046
15	dining car*	38032
16	sleeping car*	32058

*Car unoccupied at the time of the accident

Wreckage diagram (not to scale). Car numbers reflect position in consist.

In case you're not following this, here's the simple lesson: sit in Car 13! If they'll let you sit in Car 16, then sit in Car 16!

The Odds—and Improving Them

So what are the odds? Generally speaking, your odds of dying in a train crash are approximately a billion to one per mile traveled—so, for example, about five million to one for a trip from Washington, D.C., to New York. If you ride in the back of the train, however, the odds of dying on the train—unless you suffer a heart attack or drug overdose—are astronomical.

Being Struck by Lightning

Lightning fascinates us because of its destructiveness and its randomness. One moment you're standing there, minding your own business, and the next minute, a 50,000°F bolt of light carrying 300 million volts of energy comes out of the sky and smites you. Lightning is only one inch in diameter, and can strike anywhere on earth—but somehow it finds you. That's personal.*

Not surprisingly, then, the proverbial bolt from the blue has become the metaphor for long odds and random misfortune in our daily lives. "Hey, I could get struck by lightning tomorrow!" is the obligatory response to anyone questioning someone embarking on a risky adventure. And many of life's risks are now expressed in terms of whether they are more or less likely than being struck by lightning.

So, here, we have a special duty to get the odds right. Thankfully, we are aided in the endeavor by our government, namely the good folks at the National Oceanic and Atmospheric Administration (carrying the apt acronym NOAA), which is a bureau of the U.S. Department of Commerce and includes the National Weather Service. NOAA

* Unlike, say, hurricanes, lightning strikes rarely harm more than one or two people. Perhaps the greatest lightning-caused disaster was the famous explosion of the *Hindenburg* in 1937, which ended the dream of transatlantic zeppelin flights. The majority, but not unanimous, view is that the explosion was caused by an electrostatic discharge—in other words, lightning—that set off the hydrogen that filled the zeppelin. One clue: the last words of the Hindenburg's captain, Ernst Lehmann, were, "It's lightning! Jump!"

meteorologists and scientists track lightning strikes in fine, one might even say "obsessive," detail. So, too, do the soldiers of the 45th Weather Squadron of the 45th Space Wing of the Air Force's Space Command, the military equivalent of NOAA. With their help, and that of other scientists around the world, we can get a pretty firm grip on the odds.

The Odds

On average, the earth is struck by lightning one hundred times per second, for a total of 3.2 trillion strikes per year. The United States is struck approximately 31 million times per year.* But the key question is, how often does fate direct lightning to some hapless bystander?

Since 1959, local offices of the National Weather Service have kept monthly records of lightning strikes (and other weather data) and reported them to NOAA's National Climatic Data Center in Asheville, North Carolina. That Data Center produces a report, *Storm Data,* with both national and regional statistics. The most recent aggregate statistics reported by NOAA cover the thirty-five-year period 1959 to 1994.

The Data Center tracks both deaths and casualties (deaths plus injuries) from lightning. Given that almost everyone struck by lightning gets a little dinged up, the casualty data equate to the number of people actually struck by lightning. So we know that over the thirty-five-year period 1959 to 1994, an average of 363 people per year were struck by lightning, with 90 per year killed. According to Census Bureau data, the average U.S. population over this period was 209

* The number of lightning strikes is monitored by the National Lightning Detection Network. If you're having trouble passing the time, you can go to their website, *www.lightningstorm.com,* and track every lightning strike in the United States in real time.

million. Thus, the annual odds of being struck by lightning were, and presumably still are 576,000 to 1; the odds of being killed by lightning are 2.32 million to 1. (That works out to roughly one in 87,000 lightning bolts actually striking someone, with one in every 345,000 bolts killing someone.)

Improving the Odds

State by State

Florida easily leads the nation in terms of both persons struck and persons killed by lightning. This leadership role is attributable both to a large number of lightning strikes and a large population. To assess your own personal odds, however, the relevant statistic should be strikes or deaths per capita—that is, adjusted for population. Making that adjustment, we discover that the most dangerous state is actually New Mexico, followed by Wyoming, Arkansas, Florida, and Mississippi (finally at the top of some ranking!). The safest states are Alaska and Hawaii (zero deaths each), fol-

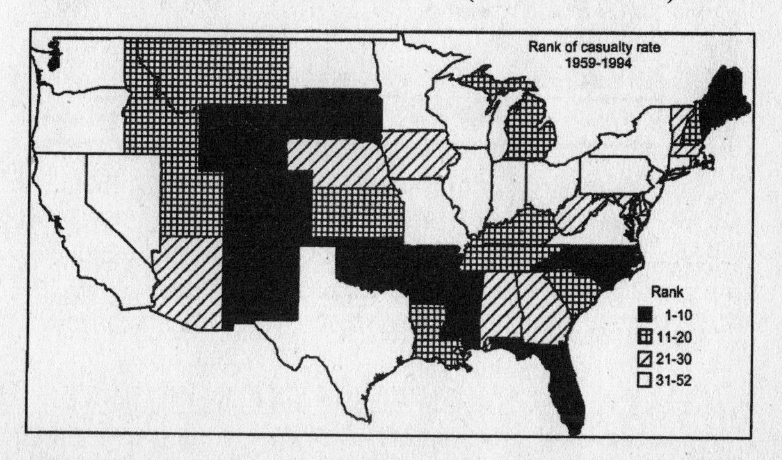

Source: NOAA

lowed by Washington, California, and Oregon. The chart on the preceding page shows the relative ranking of each state.

Basically, you'll wish to avoid the plains states and the southern states. Interestingly, while the Pacific Northwest is quite rainy, it does not produce killer thunderstorms.

Monthly Data

Your chances of being struck by lightning vary significantly by month, with the summer months the most dangerous, as the following chart shows:

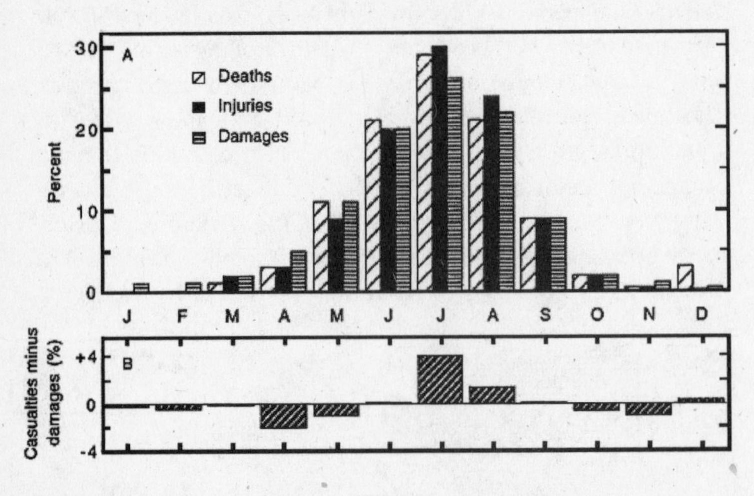

Source: NOAA

You might expect that the higher incidence of casualties and deaths in the summer is attributable to people being outdoors more frequently; we see from the chart, however, that property damage from lightning—that is, economic losses to property—also peaks in the summer and troughs in the winter. Since there's no reason to believe that buildings are any safer in winter than summer, the primary factor appears to be that violent storms, and concomitant lightning, are more prevalent in the summer months. The chart at the bottom subtracts damage rates from casualty rates to approximate how much of the reported results are attributable to human behavior, as opposed to weather patterns, and shows that behavior plays only a limited role.

Time of Day

Lightning tends to cluster at certain times of day. As thunderstorms are primarily an afternoon phenomenon, the odds of being struck at, say, 3 P.M., are about five times as great as at 9 A.M. The middle of the night is the safest time, both because there are fewer storms and because people are in bed. Here, then, behavior plays a larger role than with monthly variation.

Women v. Men

One of the primary determinants of your odds of being struck by lightning is your sex. The data (and most likely factor explaining the data) are presented graphically:

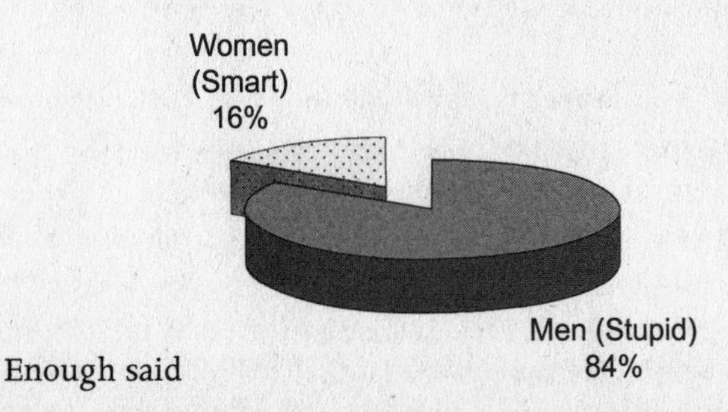

Enough said

Other Means

Lee Trevino once said that the secret to avoid being struck by lightning was holding a one-iron, because not even God can hit a one-iron. But, then again, Lee Trevino has been struck by lightning twice, so we may wish to examine alternative strategies.

The good news here is that chances of being struck by lightning steadily declined from the 1960s through the 1980s, with only a small upswing in the 1990s. Improved weather forecasting and warning systems were the major factors, but so also was better education. Death rates have declined even more markedly than injury rates, as the use of lifesaving cardiopulmonary resuscitation (CPR) has become widespread.

(You can be forgiven if you have the opposite impression, for while your individual odds have been improving, the total number of injuries and deaths has actually increased, as the population has grown. Thus, when you turn on your local news station—for which a lightning victim is right up there with someone trapped at the bottom of a well as an ideal lead story—you're more likely to see a "killer lightning" story than ever before.)

In any event, if a thunderstorm is approaching, you can decrease your odds markedly by taking a few simple steps:

1. If you are in an open field, ballpark, or playground, recognize that you are at maximal risk and promptly seek shelter. The same is true if you are in or on a body of water.
2. Avoid trees, as they are the next most dangerous place to be. They attract lightning, and their roots can conduct the electricity to you.
3. Once inside, stay away from electrical and plumbing materials, as they can conduct electricity. (Plus, if lightning blows

you off your toilet, you're certain to end up on *Cops* or *When Killer Toilets Attack!* or some other Fox show.)

4. If you can't reach shelter, get in a car. Interestingly, cars are not safe because of their rubber tires, which lightning can jump in a nanosecond, but rather because of the steel around you, which absorbs the electricity and conducts it away. So, if you have a choice between (1) a convertible, or (2) a hardtop with four flat tires, go with number 2.
5. If you are stuck in an open area, drop into a low crouch. You are best off low to the ground, and thus less likely to attract lightning. You should not lie down, however, as the ground will conduct electricity if lightning strikes nearby.

Generally speaking, if you are a woman, follow your instincts. If you are a man, note what the women are doing, and do that.

BEING POSSESSED BY SATAN

Satan's image has definitely improved of late, with the Prince of Darkness recently portrayed by the Princess of Estée Lauder, Elizabeth Hurley, in *Bedazzled*. Not to mention Al Pacino in *The Devil's Advocate*. Gone from the public consciousness are the ugly images of Linda Blair in *The Exorcist* and that creepy kid Damien in *The Omen*.

But that is not to say that Satan is not still actively pursuing mortal souls. If you harbor any doubt that you, a relatively generous and law-abiding person, could suffer demonic possession, then consider this alarming fact: *Mother Teresa was possessed by Satan*. Shortly after her death in 1997, the Archbishop of Calcutta, Henry Sebastian D'Souza, announced that he had ordered an exorcism performed on Mother Teresa shortly before she died. D'Souza said he thought the Nobel Peace Prize–winning nun possibly was being attacked by the devil, and asked a priest to exorcise it. While the exorcism was apparently not a full-blown, head-spinning affair—more akin to a tune-up than a complete engine overhaul—it nevertheless should give pause to those who believe themselves to be "above" demonic possession.

The Odds

Possession

According to exorcism expert Michael Cuneo, a sociologist at Fordham University and the author of *American Exor-*

cism, the Catholic Church now has ten full-time exorcists stationed in the United States and over three hundred in Italy.* Father Gabriele Amorth, the chief exorcist of Rome, claims to have performed over fifty thousand exorcisms.

Moreover, while most people associate exorcism with the Catholic Church, many other Christian sects perform exorcism. According to recent estimates, there are at least five hundred, and perhaps one thousand, evangelical Christian exorcism ministries in the United States alone. So the volume of possession is much greater than Catholic numbers alone would suggest.

Thus, if one calculates that there are 750 exorcists in the United States, and that they perform one exorcism per week, with two weeks' vacation, then that's 37,500 exorcisms per year. (We assume only one per week, as there is always travel involved, time to get to know the friends and family; plus time for cleanup, as exorcism can be a messy affair. Presumably, there is also a celebration afterward—think about the night Tom Arnold had when Roseanne moved out.) Assuming that there are not multiple possessions of the same person in the same year, that puts the odds of Satanic possession at approximately 7,000 to 1 for each American, per year. Unless you take the Iranian view that the Great Satan spends an inordinate amount of time in the United States, these odds should translate worldwide.

Of course, the odds of receiving an actual exorcism do vary depending on who you are. Exorcism is a Christian rite, particularly prevalent among Pentacostals, and traditionally associated with Catholics. Jews, on the other hand, do not

* As a technical matter, every Catholic baptism includes an exorcism of any evil that the child has picked up since birth. We will discount these "prophylactic" exorcisms.

believe in exorcism; their faith instead resides in expensive and prolonged psychiatry.

Exorcizing Your Demon Within

The word "exorcism" is derived from the Greek: *ek* and *horkizo,* meaning "I cause [someone] to swear." It thus is often referred to as putting a demon "on oath." The origins of exorcism are described in the New Testament. In Mark 5:1–13, Jesus entered the land of Gerasenes and met a man possessed of an unclean spirit. When the man approached Jesus, the demon inside him recognized Christ, and Christ recognized the demon. Jesus then confronted the demon and asked his name. "My name is Legion," answered the spirit, "for we are many." In response, Jesus sent the demons into a nearby herd of swine, who then jumped into the sea and drowned. (There is a rumor that they subsequently reemerged to form the Fox television network, but this has not been proven.)

While it's not hard to imagine Jesus summoning up and casting out a demon, this would not appear at first blush to be work for mere mortals. But the New Testament makes clear that others below Jesus' level can take up the cross for this purpose. In the book of Matthew (10:1), it is written, "Now, not only did Christ exorcize demons, or unclean spirits, but he gave the powers to his disciples. He gave the power against unclean spirits, to cast them out, and to heal all manner of sickness, and all manner of disease."

In the Catholic Church, the procedure for exorcism, known as the *Rituale Romanum,* proceeds in stages:[8]

- *The Presence.* The exorcist becomes aware of an alien feeling or entity, though the possessed person still appears to be responsible for his or her own actions. Through persistent questioning, the exorcist attempts to learn the identity of the demon inside. The process is not terribly different from meeting a beautiful, charming woman whom you know to have been divorced three times.

- *The Breakpoint*. The demon's Pretense evaporates, and it reveals itself. This revelation is attended by a fair amount of grunting, screaming, and bad odors. Once revealed, the demon turns on the possessed person, pointing out his or her flaws in an abusive manner. The scene is reminiscent of an unannounced visit from one's mother-in-law.
- *The Voice*. Part of the Breakpoint, the voice (of the demon) becomes "inordinately disturbing and humanly distressing babel." The exorcist must wait for the demon to have its say before proceeding to . . .
- *The Clash*. At this point, the priest chants "Rock the Casbah" and "Should I Stay or Should I Go?" Actually, this point is where the exorcist is now in direct battle with the demon, demanding that the demon reveal more information about itself, and then using this information to control and eventually destroy the demon. The exorcist is fighting to drive the demon back to Hell, where it belongs; the demon is fighting to stay in the body of the possessed.
- *The Expulsion*. Victory for the home team! The demon departs, generally with receding noises and voices, like the end of a Pink Floyd song. The possessed one is freed and proceeds immediately to *The Jerry Springer Show* to tell his or her story.

Of course, that's how it's *supposed* to go. But success is not guaranteed. Lest you have any doubts about what a tricky thing it is to perform an exorcism, consider that Pope John Paul II has performed only one exorcism during his entire papacy—on a nineteen-year-old Italian woman in 1982—*and failed*. She remains possessed to this day. The

Roman Catholic Church's chief exorcist excused the pope's failure by noting that the woman displayed "superhuman strength," but, duh, she was *possessed by Satan*. If she didn't have superhuman strength, then her electrician could have performed the damned exorcism.

But the pope's patient at least lived to fight Satan another day. Many possessed meet a grisly end at the hands of their exorcist. In 1997, a woman was stomped to death by her exorcist team in Glendale, California. Another woman was pummeled to death by a team of ministers in San Francisco in 1995. (And this is in the United States, where plaintiff's personal injury lawyers stand as a courageous force dedicated to deterring this type of conduct; be assured that if you're possessed in a less litigious nation—say, Peru—the gloves are *really* going to come off.)

Leaving aside actual death, there remains the problem of misdiagnosis. This may come as a shock, but, especially in olden times, townsfolk tended to confuse mental illness or epilepsy with demonic possession. While this trend has certainly subsided, it nonetheless persists. A Wisconsin woman successfully sued her psychiatrist in 1997, after he diagnosed her as possessed by Satan and exhibiting 126 distinct personalities, including the bride of Satan and a duck. The experience left her suicidal.[9]

Improving the Odds

So if you do think yourself possessed by Satan, how best to improve your odds of appropriate diagnosis and successful exorcism? The Pope's 0–1 record notwithstanding, you really do want to see a Catholic priest. Why? Well, they have just what you're looking for in any contractor: experience, licensing, and a manual.

The exorcism manual for the Catholic Church is named *De Exorcismus et supplicationibus quibusdam*. While the book remained set in stone for centuries after its original publication in 1614, it was updated by Pope John Paul II in 1998. (Perhaps he was just looking to revitalize the royalty stream.) The manual calls for the exorcist to be accompanied by an assistant exorcist and—thank you, plaintiff's bar—an actual doctor. So if something goes wrong, you have good old-fashioned twenty-first-century medicine to fall back on.

There's another reason to call on the Catholic Church, particularly if you are not the one possessed: they put on a darned good show. There's a lot of recited Latin, waving of crucifixes, incense, and even the use of the bones of saints—sort of like a weekend at Michael Jackson's Neverland Ranch.

With Pentacostal and other evangelical Protestant sects, on the other hand, you have no idea what you're getting. As noted above, you may be beaten to death, or told you are a duck. More typically, though, the exorcist will just scream at you a lot, or simply do laying on of hands and quiet prayer. This will bore both you and the demon inside you.

Tick list: Do's and Don'ts

1. Avoid possession in the first place—don't get involved in the kind of activities that got Mother Teresa in trouble.
2. If possessed, go Catholic.
3. Film your exorcism—you never know where reality TV is heading next.

Destruction of the Earth

You've seen the movie. Large asteroid or comet heading for the earth. Scientists (generally including one adorable, lovelorn scientist) busily scheming on how to avoid apocalypse. People running through the streets, screaming. A small dog left behind . . . But could it happen? You bet it could. And the stakes, obviously, are very large. After all, who bothers to pay the mortgage if an asteroid the size of Nebraska is looming on the horizon?

Until relatively recently, though, no one had really done any serious analysis of the odds of the earth being struck by some large object. The first such analysis was completed in 1993, when NASA finished its Spaceguard Survey Report. That report attempted to catalogue objects—mostly asteroids, but also some comets—in near-Earth orbit (known in the new trade of asteroid watching as "NEOs"). These NEOs hurtle through the galaxy in orbits that every few decades take them close enough to the earth to one day strike us. (Halley's Comet, for example, is an NEO, and Edmond Halley a pioneer in this field.)

So what are the odds of an asteroid ruining your whole day? First, we need to specify how large an asteroid we're talking about.

- *Small*—less than one hundred meters in diameter (half a football field, or the size of a five-story office building).*
These objects are extremely common.
- *Medium*—more than one hundred meters in diameter but less than one kilometer in diameter. According to NASA, there are approximately a million objects of this size in near-Earth orbit, of which about 2,250 have been identified and tracked.
- *Large*—more than one kilometer in diameter. Astronomers estimate that between 900 and 1,230 objects this size are in near-Earth orbit. A majority have now been identified and tracked.

So what are the ramifications, and the odds, of the earth being struck by each of these objects?

The Odds

Small

Consider, the events of June 30, 1908, in the Tunguska region of Siberia. An asteroid—which fortunately chose one of earth's most desolate places to alight—caused a firestorm that leveled hundreds of square miles of forest. People fifty miles away were knocked off their feet and unconscious. It brightened the night sky as far away as London. But here's the scary part: this was a relatively small asteroid—only thirty to sixty meters across—and it never even reached the earth's surface. It exploded before it got that far.

Still, because most asteroids this size don't cause such

* Apologies to American readers, but measurements in the NEO world are uniformly metric. So just remember that a meter is about the length of a yard, and a kilometer is about 70 percent the length of a mile.

significant damage, we're not even going to worry about the odds here. You'll soon see that they're the least of our problems.

Medium

The objects most likely to do major damage are asteroids of one hundred meters or greater in diameter. The odds of an object this size hitting the earth over the next one hundred years are about *5 to 1*. An asteroid in the one-hundred-meter to one-kilometer range would devastate whatever area it struck, killing every person for hundreds of miles—but probably would not imperil civilization as a whole. Scientists believe that a one-kilometer asteroid struck Southeast Asia about 750,000 years ago; we don't know the extent of the damage, but humanity obviously survived it.

Large

If an asteroid two kilometers in diameter were to strike the earth, the impact would be akin to 5 million Hiroshima-sized nuclear bombs, extinguishing most life on Earth, including our own species. Indeed, scientists now believe that it was an asteroid about twenty kilometers in diameter that struck the earth 65 million years ago in Mexico's Yucatán Peninsula, extinguishing the dinosaurs.

According to the NASA Ames Research Center, an asteroid of two kilometers or more in size collides with the earth once or twice per million years. That means that your odds of dying as the result of a major asteroid collision are only 20,000 to 1.[10]

If that isn't sufficiently jarring to you, consider that your chances of dying by asteroid are:

- 3 times greater than dying in a bus or train crash
- 3 times greater than dying in an earthquake

- 25 times greater than dying of a snakebite
- 250 times greater than dying from a shark attack.

Improving the Odds

In this case, improving the odds means having someone with the right equipment and know-how: (1) track objects in near-Earth orbit, and (2) figure out what the heck to do if one of them eventually ends up on a collision course with Earth. More time and money is being spent on this mission than you probably appreciate.

The Spaceguard Survey that began in 1993 continues apace, and now is being carried out jointly by NASA and the U.S. Air Force. They are under a Congressional mandate to detect 90 percent of the objects one kilometer or larger by 2008. They sponsor events like 2002's NASA Workshop on Scientific Requirements for Mitigation of Hazardous Comets and Asteroids. Eventually, their work will turn to determining how best to destroy or deflect objects that present a clear and future danger.

Let's all wish them luck.

Destruction of the Universe

Even if you were convinced that some large asteroid were on an Earth-dooming trajectory, there would still be reason for hope. You might console yourself by saying, "Hey, so what if all life on Earth is going to be destroyed sometime in the next thousand years or so. By then, my multi-great-grandchildren will be exploring space, and have found a new planet to settle on."

Well, that assumes, of course, that the entire universe is not slated for destruction in a rerun of the Big Bang. Whether that is in our (distant) future essentially boils down to whether the universe is (1) contracting, (2) expanding at a decelerating rate, or (3) expanding at an accelerating rate. And here for once we have some good news. Well, relatively good news.

Here's a quick refresher for those who don't quite understand the Big Bang theory. In a nutshell, according to NASA's official version:

> The Big Bang did not occur at a single point in space as an "explosion." It is better thought of as the simultaneous appearance of space everywhere in the universe. That region of space that is within our present horizon was indeed no bigger than a point in the past. Nevertheless, if all of space both inside and outside our horizon is infinite now, it was born infinite. If it is closed and finite, then it was born with zero volume and grew from that. In neither case is there a "center of expansion"—a point from which the universe is expanding away.

If you conceptualize the universe as shaped like a ball—as we'll see, it really isn't, but at least it's a concept we can understand—then according to NASA "the radius of the ball grows as the universe expands, but all points on the surface of the ball (the universe) recede from each other in an identical fashion. The interior of the ball should not be regarded as part of the spatial universe in this analogy. In fact, the interior of the ball should be regarded as the past (its center is the "beginning," and at the exterior of the ball is the future, which the ball is expanding into)." Got that?

Since the time the Big Bang theory gained wide acceptance, cosmologists have wondered about the trajectory, and ultimately the fate, of the universe produced by that Bang. This question is one that haunted no less a mind than Albert Einstein's. In his Theory of Relativity, Einstein assumed that the universe was static, while his calculations showed that any universe with matter should be expanding or contracting. He tried to fix that by adding a "fudge factor" called the cosmological constant. When, a few years later, Hubble showed that the universe was expanding, Einstein called the cosmological constant his "greatest blunder."

With all this as background, one could argue with considerable justification that February 11, 2003, was the most important day in human history. Imagine a day when it is publicly announced that we have photographs that:

- tell us the exact age of the universe (13.7 billion years);
- further confirm that the theory of a Big Bang is actually fact;
- reveal that the first stars formed 200 million years after the Big Bang;
- inform us that the universe is not either positively curved (like a ball) or negatively curved (like a saddle) as many had

theorized, but actually flat (meaning, yes, that the earth is not flat but the universe is)*;

- confirm that in the first 10^{-35} seconds of its existence,** the Universe got 10^{50} times bigger; this is called the inflation rate by cosmologists, and certainly puts to shame the earthly version;
- confirm that there are other forces in the universe besides the matter that makes up our solar system—in fact, confirm that the universe is 73 percent "dark energy," 23 percent cold dark matter (heretofore undiscovered massive particles), and only 4 percent atomic matter (that's us, and everything we see);
- inform us that, because the force of the dark energy is sufficiently great to overcome the gravitational pull of the atomic matter, the universe is expanding at an accelerating rate, and that there will never be a second Big Bang, or re-Bang (and that Einstein didn't blunder after all, because adding his constant to an expanding universe can accelerate the expansion).

February 11, 2003, was that day. Of course, you can be forgiven if you missed this news entirely—if you don't remember where you were and what you were doing when the photographs were released. The news media relegated it to a one-day story. After all, it lacked the significance of Robert Blake's murder trial or big snowstorms on the East Coast. How can blurry pictures documenting the history and predicting the future of our universe possibly expect to keep the attention of Rather, Brokaw, and Jennings for more than a day

* If you can even begin to conceptualize what it means to have a flat universe, you need to be reading more high-brow literature than this book. And by the way, according to scientists, it's actually flat "with adiabatic Gaussian fluctuations on a scale-invariant spectrum." Really.

** To get a sense of how short a period of time that is, recognize that 10^{-6} seconds is a millionth of a second, and 10^{-9} is a billionth of a second.

when there are beautiful pictures of people shoveling snow, and the anchors get to keep repeating the ratings-grabbing phrase "killer storm"?

In any event, February 11 was the day that NASA went public with the product of the Wilkinson Microwave Anisotropy Probe (or WMAP), a dinky little $145 million thing that was launched with little fanfare in June 2001 and tasked with measuring cosmic microwave background radiation. The most tangible product of the WMAP satellite was the following picture, which shows the universe 379,000 years after the Big Bang:

While 379,000 years may seem like a long time after the event, consider that given how long ago the Big Bang occurred, this is the equivalent of a photograph of an eighty-year-old man taken on the day of his birth. Think of it another way: 380,000 years post-Bang was the first moment that the universe became "photo ready"; put another way, the formation of the stars was the flashbulb in the dark room of the universe that let our cameras take a picture.

Note how darn cool it is that we are now taking photographs of the *past*. Of course, that's because these images are from billions of light-years away, and since they travel at "only" the speed of light, they are just now reaching us. Conversely, we will not be able to take a photograph of what the other end of our universe looks like *now* until billions of years have passed. Effectively, we can *only* photograph the universe's distant past.

The Odds

As for our bottom line, the WMAP photographs confirm that the universe will continue expanding and not re-Bang. In effect, there are no odds: there is no theory under which we re-Bang. Not even London bookmakers would take odds. (Actually, they might, since they'd get to hold your stake for an exceedingly long time.)

Of course, there is some bad news as well. As the dark energy forces the universe to continue expanding, there will be less atomic matter over a far greater space. What that means is that the universe in fact will die, but (choose your poetic cliché) from ice rather than fire, or with a whimper rather than a bang.

But you're talking 10^{100} years here. So, until you hear different, keep paying the mortgage.

PART 4

THE SPORTING LIFE

Catching a Ball at a Major League Game

Time was, if someone said "catching a ball at a major league game," you would have pictured an eight-year-old with a mitt, wearing a two-sizes-too-large team jacket and waiting patiently. Now, unfortunately, the most common image is of Alex Popov and Patrick Hayashi.

Popov is the man who initially caught Barry Bonds's record-breaking seventy-second home run in October 2001, and Hayashi is the man who somehow ended up with possession of the ball, after a rather considerable scuffle. Popov said he caught the ball fair and square; Hayashi contended that Popov should have held onto it a bit tighter. Litigation resulted, and the ball was placed in a safety-deposit box. Finally, the two men talked, and a year and a half later, Popov told the press, "We have come to an agreement that the best thing to do is get back to what this is all about, which is that this is a historical moment in baseball history that we are both a part of." So they agreed to . . . sell the ball and split the money, which is, after all, what it was "all about." (If you thought he meant they were going to give the ball to the Hall of Fame, or the man who hit it, or a charitable foundation, then please reboard your ship and return to whatever planet you came from.) The ball disappointed its co-auctioneers by bringing a mere $345,000 (about equal to their legal fees).

That disappointment came in relation to the ball that constituted Mark McGuire's historic seventieth home run in 1998—which fetched $3 million (from the same buyer). The value of that ball presumably dipped quite a bit after Bonds hit number 71 and number 72. But, in any event, times have certainly changed since 1976, when Richard Arndt, a groundskeeper for the Milwaukee Brewers, caught Hank Aaron's 755th and final home-run ball. He offered to give the ball to Aaron, but wanted to do so personally, the next day. The Brewers fired him for being too pushy, and deducted $5, the price of the ball, from his last paycheck. Those were the days!

But whether you hope to get rich or simply put a smile on the face of the boy or girl you took to the ballpark, catching a ball is always a fortunate occurrence. So what are the odds?

The Odds

The basic odds of catching a ball at any major league game can be derived by dividing (1) the number of fans at the game by (2) the number of balls that end up in the seats.

The ball. We'll start with the latter part of the problem, as it is easier to figure.

- Surprisingly, the average "life" of a major league baseball is only six pitches—that is, a batter will hit either a foul ball or a home run every six pitches. If you consider that the average at-bat consists of just fewer than six pitches, then that means about one foul ball or home run per at bat.
- We also know that a major league game averages around 300 pitches (150 per team), so the average number of "lost" balls equals 300/6 (50).

So, in the average game, you're going to have fifty bites at the apple.*

The competition. But you're going to have to fight your fellow fans for your icon. In 2002, the average major league baseball game drew 28,158 fans.

The basic odds. So assuming for the moment that all fans have an equal shot at catching a ball, your odds at the average game are 563 to 1. If you're a season ticket holder (attending all eighty-one of your team's home games), then your odds are 7 to 1.

Not all fans are created equal. The true odds will depend greatly on where you're sitting—both at which ballpark, and where in that ballpark. Why? Well, while the average game drew 28,158, attendance varies significantly by ballpark. Assuming that the number of fouls and home runs is constant,** then your odds of catching a ball are four times as great in some parks as in others.

* We'll disregard the balls that go into the screen behind the plate and are reused.

** Actually, one might assume that attendance would tend to rise with home run totals, thereby leveling off the odds a bit. But even if that were so, the great majority of ball-catching opportunities come on foul balls, which do not so vary. Moreover, attendance also tends to rise with winning, and a strong pitching staff, which is a constant among good teams, tends to give up fewer opposition home runs.

Major League Home Attendance—2002

Rank	Team	Games	Total	Avg.	Pct.
1	Seattle	81	3,542,938	43,739	93 percent
2	NY Yankees	80	3,465,807	43,322	79 percent
3	San Francisco	81	3,253,203	40,163	98 percent
4	Arizona	81	3,198,977	39,493	88 percent
5	Los Angeles	81	3,131,255	38,657	69 percent
6	St. Louis	81	3,011,756	37,182	75 percent
7	NY Mets	78	2,804,838	35,959	65 percent
8	Chicago Cubs	78	2,693,096	34,526	89 percent
9	Colorado	81	2,737,838	33,800	67 percent
10	Baltimore	81	2,682,439	33,116	68 percent
11	Boston	81	2,650,862	32,726	97 percent
12	Cleveland	81	2,616,940	32,307	75 percent
13	Atlanta	81	2,603,484	32,141	64 percent
14	Houston	81	2,517,357	31,078	74 percent
15	Texas	80	2,352,397	29,404	60 percent
16	Anaheim	81	2,305,547	28,463	63 percent
17	San Diego	81	2,220,601	27,414	41 percent
18	Oakland	81	2,169,811	26,787	61 percent
19	Milwaukee	81	1,969,153	24,310	57 percent
20	Minnesota	81	1,924,473	23,758	49 percent
21	Cincinnati	80	1,855,787	23,197	58 percent
22	Pittsburgh	79	1,784,988	22,594	59 percent
23	Chicago Sox	81	1,676,911	20,702	47 percent
24	Philadelphia	79	1,618,467	20,486	33 percent
25	Toronto	81	1,637,900	20,220	40 percent
26	Detroit	80	1,503,623	18,795	47 percent
27	Kansas City	77	1,323,036	17,182	42 percent
28	Tampa Bay	81	1,065,742	13,157	29 percent
29	Florida	81	813,118	10,038	24 percent
30	Montreal	81	812,045	10,025	22 percent
	AVG	80	2,264,813	28,158	61 percent

Note that odds vary not just by attendance, though, but also by density. Compare, for example, the Chicago White Sox and Philadelphia Phillies, which had practically identical attendance figures, with resulting pro forma odds of around 400 to 1. But note that Philadelphia has a much larger ballpark, which is on average only 33 percent full, whereas the

White Sox show 47 percent of capacity. That means that if you're an energetic teenager, you've got more room to roam, and thus better odds, in Philadelphia.

Improving the Odds

Generally, you can't control the city where you will attend games, or your physical agility. Thus, improving your odds of catching a ball is going to consist primarily of carefully choosing where you sit in the ballpark.

First question: should you sit in the outfield bleachers and go for the home run ball, or stick to the first or third base line and look for foul balls? Well, in 2002, there were 5,059 home runs hit in Major League Baseball. Given that thirty teams played a total of 4,852 games, that means that on average only one to two of the fifty balls available for catching were home runs. Furthermore, at certain ballparks, most notably Wrigley Field in Chicago, fans are expected to throw back home runs hit by the opposing team. Since no such rule applies to opposition foul balls, the attractiveness of the outfield diminishes further.* So, generally we're looking at foul territory.

Not all foul territory is created equal, however. A net protects fans sitting behind the plate from getting struck by a foul ball, so that area is out. The upper deck beyond the bases is simply out of range. So, you really want to be on a lower deck, along the baselines.

You may also wish to consider the configuration of the ballpark you're visiting. While going for home runs is generally against the odds, for example, the Ballpark in Arlington

* Note that there are numerous other attractions to sitting in the bleachers at Wrigley Field; here, however, the analysis is focused only on ball catching.

(home of the Texas Rangers) has a notoriously short right field wall.

Once you've found a seat, are there any other tactics you may wish to employ? Well, crack researcher Patrick Ferguson spoke with Ron Falstaff, an usher at Orioles Park at Camden Yards, home of the Baltimore Orioles. Mr. Falstaff offers a few tips:

- First, most foul balls are caught on the bounce or ricochet, so don't take your eyes off the ball just because it has flown over your head.
- Second, try to take your glove, lest one of those bounces come off your broken finger.
- Third, the exits and walkways get their share of balls, which are generally caught by the most eager and observant fan in the vicinity, so keep your eyes open during your trips to and from your seat.

Play ball!

Reaching the Summit of Mount Everest

Human beings have dreamed of climbing Mount Everest since 1852, when the Great Trigonometric Survey of India determined that the peak was in fact the world's highest. Until then, it had been designated only as Peak XV, and was unknown in the West. It was renamed, appropriately enough, for the British surveyor who rescued it from obscurity.

Even after its altitudinal coming out party, however, Mount Everest remained off limits to climbers. While the mountain extends into Tibet, Nepal, and India, most climbers believed that the only possible route to the summit was through Tibet, but Tibet initially refused to allow visitors. Only in 1921 did attempts to reach the summit begin.

These attempts did not go particularly well. The British made it a matter of national pride that someone from the Empire reach the summit first. The best known climber of the day was George Mallory of England, and he made three attempts at the summit, beginning in 1921. Mallory, who is best remembered for saying he wished to climb Everest "because it is there," did not attempt a fourth climb because he was, well, dead—the victim of an unexpected storm during his third push for the summit in 1924. There was considerable speculation that Mallory had actually reached the summit before expiring, but no one really knew. (His body was found seventy-five years later, and while it was near the summit, there was no evidence to show that he died on the way down as opposed to the way up.)

Ten more expeditions followed Mallory, and until 1953 the record stood at Everest 10, Mountaineers 0, with fifteen dead. And in 1950, Tibet was invaded by China and closed to mountaineers for the next thirty years; by then, however, an alternate route through Nepal had been established. Using that route, New Zealander Edmund Hillary and his porter, Sherpa Tenzing Norgay of India, finally reached the summit.* They have never revealed who was actually first (Buzz Aldrin should have been so lucky, having to sit on his keister in the lunar module listening to Neil Armstrong let loose with "One small step for man, one giant leap for mankind.") This is actually extremely sporting of Hillary, who was a paying customer and thus presumably entitled either to go first, or to go second and claim to have gone first. There were some nice, quaint traditions in the old British Empire. . . .

While originally a British pursuit, the dream of Everest has over time proved to be a remarkably egalitarian one. Everest has been climbed by:

- the old (sixty-one-year-old Georgian** Lev Sarsikov)
- the young (fifteen-year-old Nepalese student Temba Tsheri)

* Three trivia notes. First, Hillary is actually still alive. (Sadly, no one under the age of thirty even knows who he is—you can check—which is unfortunate given not only his heroic deeds but also the dignity with which he has left them behind—as opposed to engaging in a lifelong Norman Schwarzkopfesque victory tour of appearances and paid-for speeches.) Second, "Sherpa," in the name "Sherpa Tenzing Norgay," is not an adjective; Norgay's name—like everyone's in his community—actually is "Sherpa," which is appropriate given that that is the dominant profession. (Imagine such a system in the United States—say, if a typical name in Kansas was Farmer Bobby Jones, or in Hawaii was Surfer David Ho.) Third, if you were wondering who the most famous beekeeper in world history is, the answer is Sir Edmund Hillary.

** That's Georgia, the country that used to be a Soviet republic. There is no one in Atlanta named Lev Sarsikov.

- the blind (American Erik Weihenmayer)
- the heroic and traditional (Reinhold Messner, alone and without supplemental oxygen)
- the tacky and unsporting (a Chinese army team of 410, which managed to get nine climbers to the top in 1975)
- men and women (Junko Tabei of Japan, being the first woman, in 1975)
- fathers and sons (*père et fils* Roche, of France, being the first, in 1990)

All of these feats, though, were merely prelude to the big event of May 21, 2000, when Byron Smith became the first Ford dealer to reach the summit, declaring, "I can't go any further, I'm on top of the world." (Presumably, he meant "farther.") In any event, like any good Ford dealer, he established a website in sale-a-bration of the event; learn more about this Ford explorer's expedition at *www.vulcanford.ca*.

The Odds

So what are your odds? Between 1921 and 2000, Mount Everest was climbed by more than 1,300 people from twenty countries. Unfortunately, more than 160 climbers died in the attempt. (One bummer: if you die on Everest, no one brings your body down, as it takes too much time, energy, and money. On the bright side, you never decompose. Thus, when George Mallory's grandson climbed Everest in 1995, he got to spend some quality time with his late grandfather.)

No official records are kept on how many climbers *attempt* to reach the summit each year, but a check with some Everest guide sites yields a rule of thumb that about 20 percent of paying clients make it to the top. Thus, your odds of reaching the summit are about 4 to 1; your odds of dying in the attempt are about 40 to 1. Put another way, if you are part of a group of forty climbers, the odds are that eight of you will make it, thirty-one will head home failures, and one will die. So here's some advice: instead of taking your college buddies to Everest, how about golf in Scotland instead?

Improving the Odds

The Smith ascension is an example of a growing trend in the 1980s and 1990s, when tour guides promised that they could get any fit individual to the top, and often proved their point. Furthermore, whereas George Mallory attempted the summit dressed only in cotton and wool, along with primitive oxygen supplies, today's climbers can insulate themselves with goose down, GORE-TEX, and far more sophisticated breathing equipment.

The result has been record numbers of successful ascents (182 in 2001 and 159 in 2002) and deaths (15 in 2000). But if you're bound and determined to go, an experienced guide with a low casualty rate is the best option.

Hitting a Hole in One

A hole in one is one of life's most celebrated triumphs over odds, because it requires a unique blend of skill and luck. The ball must be struck reasonably well, to land somewhere on or very near the green. But the golfer knows that he or she could strike thousands of balls perfectly, and still never get a hole in one, and that the same is true for even the best golfers in the world. A hole in one represents the intervention of great fortune in the midst of sport.

The Odds

But what are the odds? We begin with odds that are the easiest to measure—the odds that a PGA Tour player will make a hole in one. Because the PGA Tour documents its play meticulously, we have the records to know the odds fairly exactly. Calculating the odds for a given year involves first determining how many rounds were played and multiplying the number of rounds by the average number of Par 3s per round; the latter we can assume to be four, as that is standard. This product gives us the number of Par 3s played on the PGA tour each year. The next question is how many golfers play those holes each round. With the help of the kind folks at the PGA Tour, we estimate that number as 110.*

* The general PGA Tour event has a field of 144 players until daylight savings time begins, and 156 players after daylight savings for an average of 150. There

Multiply the number of Par 3s played by the number of golfers playing them, and you get the number of chances at a hole in one. Divide that number by the actual number of holes in one, and you have the odds.

Odds of Hole in One—PGA Tour

Year	Rounds	Par 3s	Avg. players/ round	Holes in one	Odds
1993	172	688	110	25	3,027 to 1
1994	172	688	110	44	1,720 to 1
1995	176	704	110	35	2,212 to 1
1996	180	720	110	40	1,980 to 1
1997	180	720	110	32	2,475 to 1
1998	180	720	110	32	2,475 to 1
1999	188	752	110	30	2,757 to 1
2000	196	784	110	29	2,973 to 1
2001	196	784	110	28	3,080 to 1
2002	196	784	110	39	2,211 to 1
Avg.	184	734	110	33	2,491 to 1

Source: PGA Tour

So over the past ten years, the odds of the average PGA golfer making a hole in one have been 2,491 to 1. While the sample size is not very large, it appears that better golfers tend to get more than their fair share (if for no other reason than that by making the cut, they double their number of chances). Thus, over the past ten years, Tiger Woods has had two,

are numerous invitation tournaments, though, with unique field size: for example, the Mercedes Championship includes only 36 players, while the ATT Pebble Beach includes 180. Generally, though, most of the invitational tournaments are smaller than average, and thereby bring the average field down to around 145. On the PGA Tour, though, there is a "cut" made after two rounds. The cut is generally made at seventieth place, with those tied for seventieth included; that rule leaves an average of 75 players for the weekend. So, with 145 players competing for the first two rounds, and 75 competing for the last, the average number of players per round is 110.

Davis Love III has had one, Phil Mickelson has had three, and Ernie Els has had one. But fate smiles on the lesser knowns as well: Jonathan Kaye had two in 1999; Mike Brisky has had three, and two in two weeks in 1999; and both Bob Tway and Glenn Day each had two at a single tournament (the Memorial and the Greater Hartford Open, respectively) in 1994. As Dr. Seuss would say, "Way to play, Day and Tway!"

But, as the PGA Tour is fond of saying, "These guys are good." In fact, much better than you, the average hacker. So unless we knew exactly how much better, we can't really calculate your odds from their odds. Fortunately, though, we have two other methods.

Method one. One thing is for sure about holes in one: when golfers get one, they tend to let people know about it. By tradition, they are obligated to buy a round of drinks for anyone in the clubhouse, but that is a small price to pay. Getting a hole in one is like having a baby or getting engaged—events that must be instantly communicated to everyone you know, and a few you don't.

Among those hearing the news is the pro shop of the lucky course, where the golfer initially spills his or her guts and begins a desperate hunt for souvenirs to mark the Big Day—certificates, items on which to mount the lucky ball, even dioramas of the lucky Par 3s. What that means, though, is that clubs generally know how many holes in one their course is producing. So we contacted a few clubs in a random survey to learn about their hole-in-one rates.

- Pebble Beach Golf Links is often rated as the best golf course in the world, and would certainly rank as the best place to get a hole in one, since they give you the flag from the hole and a certificate, free, to celebrate your achievement. According to Sergio Castillo, first assistant golf professional at the course, they average 62,000 rounds per year, and last year had thirty-three holes in one. With four Par 3s on the course, that makes the odds about 7,500 to 1. But given the difficulty and expense of Pebble Beach, they probably tend to draw very good golfers, so these odds may not hold true for the general golfing public.*

- Wintergreen is a resort in the Blue Ridge Mountains of Virginia with forty-five holes of golf designed by famous guys. According to Head Professional Lance Reynolds, the shorter mountain course (with five par threes) yields an average of seven holes in one per year over 22,500 rounds, making the odds 16,000 to 1. The twenty-seven holes in the valley include six par threes, which are longer; they yield an average of five holes in one over 39,000 rounds. So the odds are around 47,000 to 1.
- East Potomac Park in Washington, D.C., is one of the nation's busiest municipal golf courses, attracting 120,000 rounds per year over two eighteen-hole courses. According to Mike Byrd, the manager at the course, only three holes in one are generally reported in a given year. Given that each course includes six par threes, that implies odds of 240,000 to 1. But the course is a busy place, and doesn't sell those valuable hole-in-one trinkets, so there may be significant underreporting.

While these numbers get us a ballpark estimate, we see a lot of variation, and need to do better. And we will!

Method two. Fortunately, we have a group of people

* Note that while the odds at Pebble Beach are only three times as great as on the PGA Tour, this does *not* mean that the professionals are only three times more likely to get a hole in one as the average player at Pebble Beach. The Par 3s on the Tour (including the stop at Pebble Beach) are set up considerably longer than they are for amateur play.

whose business it is to know exactly what the odds of getting a hole in one are. Those are the insurance companies who insure against the "risk" of a hole in one. What risk? Well, for example, local charities will allow golfers approaching a par three to pay $10 for the chance at a $100,000 prize if they make a hole in one. The charity insures this risk with a property-and-casualty insurance company, thereby allowing it to make relatively predictable, risk-free profit on the event (the money collected less the premium paid). Even some PGA Tour sponsors, who give away an automobile or $1 million if a professional makes a hole in one, will decide to go to an insurer rather than self insuring and risking having to pay out themselves. Because insurance companies have their money on the line every time the ball is struck, they keep pretty good track of the odds.

For the voice of experience, we turned to Doug Burkert, who is the president of the National Hole in One Association, an insurer in this field. Mr. Burkert has been insuring against holes in one for over twenty years. According to Mr. Burkert, the three main factors in setting an insurance premium for a hole in one contest are the prize money to be awarded, the number of golfers, and the yardage of the hole. The sex of the golfers is also considered indirectly, as women are allowed to play from their regular tee, which is invariably closer to the hole, at the regular premium price. If a majority of the group was female, then they would be charged a lower premium for any given distance.

There are a variety of other, less tangible factors that affect the odds:

- A prevailing wind could make a hole predictably longer or shorter than its stated yardage.
- Pin position on the green could make the shot harder (if

near the fringe, or on a ledge) or easier (if in the center, or an area where the cut of the green tends to feed the ball).

- Water close to the hole makes the shot harder, as lucky bounces don't happen.
- Contests at country clubs are more likely to yield a hole in one than those at municipal courses, as the golfers tend to be better.
- If the contestants are members of a country club where the contest is being held, there is a greater likelihood of a hole in one, as members know how the green rolls.

Interestingly, though, Mr. Burkert tells us that premiums are not adjusted for these factors, as they are not significant enough to warrant an in-person inspection, and there is no reliable mechanism for reporting how these factors might affect the odds on a given hole. (Similarly, while it would certainly make economic sense to know the handicaps, or skill level, of the participating golfers before underwriting the policy, doing so is impractical, not only logistically but also because many golfers don't carry a handicap, or carry a misleading one. Thus, most insurers only distinguish for underwriting purposes between professionals and amateurs.) Still, economically, the uncertainty caused by these elements increases the risk to the underwriter, so they are priced into the premium.

So what are the odds? According to Mr. Burkert, most contests are conducted at distances of 150 to 175 yards, and the odds at that distance have traditionally been 12,600 to 1. In the last couple of years, Mr. Burkert suspects that improved club and ball technology have lowered those odds a bit, to about 12,000 to 1. Distance does matter, though. Mr. Burkert generally won't price hole-in-one insurance if the hole is shorter than 135 yards. He reckons that at that dis-

July 4, 1997, the author beats the odds.

tance, the odds are probably about 10,500 to 1. At 200 yards, the odds rise to 14,000 to 15,000 to 1.

By the way, in case you were wondering, Mr. Burkert has never had a hole in one, though he'd really like to.

Just to be sure, we checked with another insurance

underwriter—though this one preferred not to be named. The odds were a bit longer than with Mr. Burkert: 20,000 to 1 at 200 yards; 15,000 to 1 at 175 yards; and 13,000 to 1 at 150 yards. Interestingly, this company had actually underwritten contests down to 120 yards, but here the odds dropped appreciably, to only 7,500 to 1.

Lesson: join the country club with the shortest Par 3s.

Bowling a Perfect Game

Here's a little SAT question for you: NFL football is to stock car racing as golf is to ______. Well, the answer, as you've most likely deduced from the title of this chapter, is bowling. But why? Given its saturation coverage, most people assume that the NFL is the most popular spectator sport in America, but NASCAR trounces it by every objective measure. Similarly, innumerable stories about the "Tiger Woods phenomenon" proclaim that America in now a nation of golfers, but according to the National Sporting Goods Association, there are 40 million bowlers and only 27 million golfers in the United States.

So why do we see the PGA and not the PBA on television every weekend? Well, maybe it has something to do with the fact that the folks who run our television networks (and magazines and newspapers) tend be well-heeled, country club types. People like fancy-pants announcer Jim Nantz (college golfer), who now sits in the Butler cabin at Augusta asking Masters participants whispered questions like, "Do you feel as honored as I do to be sitting in the Butler cabin?" Try to picture Jim Nantz sitting on a stool in the game/interview room at the Shirley-Bowl, shouting over the Ms. Pac Man game, "So, Roy, what were you thinkin' on that eight-ten split in the ninth frame?"

Surprisingly, given the disdain of Mr. Nantz and his fellow elitists, bowling has a remarkably rich history.

According to the American Bowling Congress, the origins of bowling have been traced back to Egypt circa 5200 B.C. Bowling with pins originated in Germany, and with no less a figure than Martin Luther helping to shape the rules of the European game. (He set the number of pins at nine, though a subsequent "reformation" of the rules added a tenth.)

The Odds

So having covered the odds of a hole in one, it seems only fair to examine the odds of its middle-class equivalent, the perfect game. (Note that the upper class need only make one good shot to be celebrated in the local newspaper; the middle class, as in so many things, must be perfect all day long.)

A perfect game requires twelve consecutive strikes—one in each of the first nine frames and three in the tenth. Your odds of bowling a perfect game obviously depend on your bowling skill. The good news is that those odds have been growing dramatically shorter.

Set out below are perfect game data for league bowlers—that is, those who participate in leagues accredited by the American Bowling Congress. The ABC is the NCAA of league bowling.

Odds of Bowling a Perfect Game

Year	Bowlers	# 300s	300 Odds
	(est. until '95–'96)		*(100 games/year)*
1895–1907	Na	0	
1907–08	6,640	1	
1908–09	6,640	2	
1909–10	**6,785**	**1**	**650,000 to 1**
1910–11	7,270	4	
1911–12	6,370	1	
1912–13	7,690	1	
1913–14	8,900	3	
1914–15	10,520	3	
1915–16	16,440	3	
1916–17	16,740	4	
1917–18	15,760	5	
1918–19	13,620	5	
1919–20	**25,885**	**5**	**517,700 to 1**
1920–21	24,350	4	
1921–22	30,000	2	
1922–23	58,000	10	
1923–24	51,000	17	
1924–25	64,000	21	
1925–26	76,000	25	
1926–27	94,000	30	
1927–28	110,000	53	
1928–29	139,000	58	
1929–30	**210,000**	**84**	**250,000 to 1**
1930–31	215,000	120	
1931–32	187,000	135	
1932–33	140,000	143	
1933–34	158,000	245	
1934–35	203,000	168	
1935–36	351,000	192	
1936–37	307,000	200	
1937–38	446,000	229	
1938–39	483,000	237	
1939–40	**602,000**	**284**	**211,972 to 1**
1940–41	746,000	235	
1941–42	875,000	225	
1942–43	694,000	137	
1943–44	697,000	136	
1944–45	795,000	109	
1945–46	810,000	81	
1946–47	1,105,000	133	
1947–48	1,259,000	170	
1948–49	1,368,000	167	
1949–50	**1,417,000**	**219**	**647,032 to 1**
1950–51	1,430,000	206	
1951–52	1,482,000	225	
1952–53	1,569,000	198	
1953–54	1,651,000	242	
1954–55	1,741,000	232	
1955–56	**1,929,000**	**270**	*(100 games/year)*
1956–57	2,225,000	302	
1957–58	2,500,000	347	
1958–59	3,000,000	426	
1959–60	**3,500,000**	**582**	**601,375 to 1**
1960–61	4,000,000	636	
1961–62	4,275,000	648	
1962–63	4,500,000	790	
1963–64	4,575,000	829	
1964–65	4,550,000	769	
1965–66	4,375,000	783	
1966–67	4,250,000	761	
1967–68	4,200,000	815	
1968–69	4,150,000	905	
1969–70	**4,100,000**	**854**	**480,094 to 1**
1970–71	4,000,000	985	
1971–72	4,100,000	1,163	
1972–73	4,150,000	1,178	
1973–74	4,200,000	1,377	
1974–75	4,300,000	1,606	
1975–76	4,500,000	1,913	
1976–77	4,583,000	1,375	
1977–78	4,727,000	1,912	
1978–78	4,777,000	3,426	
1979–80	**4,799,195**	**5,373**	**89,321 to 1**
1980–81	4,755,756	5,549	
1981–82	4,685,036	5,949	
1982–83	4,556,907	6,030	
1983–84	3,791,081	6,447	
1984–85	3,656,928	6,978	
1985–86	3,624,575	6,467	
1986–87	3,424,205	8,368	
1987–88	3,313,491	9,901	
1988–89	3,165,471	10,733	
1989–90	**3,036,907**	**12,766**	**23,789 to 1**
1990–91	2,922,829	14,192	20,595 to 1
1991–92	2,712,987	14,889	18,221 to 1
1992–93	2,576,809	20,542	12,544 to 1
1993–94	2,454,742	25,387	9,669 to 1
1994–95	2,370,190	29,032	8,164 to 1
1995–96	2,261,469	30,630	7,383 to 1
1996–97	2,135,126	33,276	6,416 to 1
1997–98	2,026,684	34,217	5,923 to 1
1998–99	1,936,648	34,470	5,618 to 1
1999–00	**1,866,023**	**39,470**	**4,728 to 1**
2000–01	1,767,096	41,303	4,278 to 1
2001–02	1,694,248	42,163	4,018 to 1

Source: American Bowling Congress

As you see, the odds of bowling a perfect game have improved exponentially over the past century. Fifty years ago, there was only one perfect game bowled each year for every 7,924 league bowlers; now, it's one per 40 bowlers. Assuming the average league member bowls one hundred games per year, that's odds of around 4,000 to 1 in each game—down from 89,000 to 1 only twenty years ago and 640,000 to 1 fifty years ago. As Jeff Carter, who has bowled a record seventy-two perfect games, told *USA Today,* "For my first one in 1986, I got phone calls from all over. It was in the newspaper. Now, it doesn't even get mentioned over the P.A. system in the bowling center."

The investigative journalists at *USA Today*—the only national newspaper with the guts to probe "300-gate"—identified several factors that may account for this trend: the balls, the pins, and the lanes. Strangely overlooked, though, were the factors that have largely accounted for the proliferation of records in other sports—superior training and conditioning of the athlete, frequently assisted with performance-enhancing drugs, leading to improved strength and speed. Could it be that today's bowlers have switched from regular to "lite" beer? Could it be that the spare tires around their waists are no longer rolls of fat, but steroid-toughened ballast? Unfortunately, until the Pulitzer Prize committees start to recognize what true investigative journalism looks like, we are unlikely to find out.

Clearly, though, heavier bowling balls have played a role. More importantly, they're ideally weighted to hit the all-important "pocket"—the space between the head pin and the one behind it (to the right for right handers and to the left for left handers). For maximum effect, a bowling ball should hit the pocket after veering in six degrees from the side, with the ball continuing into the middle of the pins to do more damage. To achieve that goal, the ball needs to hook. But how to get a dependable hook?

Well, you can practice until your fingers bleed, or you can help that hook by buying one of the modern off-center weighted balls. If you're willing to shell out $242 for Ebonite's "TPC Warrior," for example, you can get a "unique totally asymmetric core surrounded by the pearlized Big Wheel reactive resin." Alternatively, you might consider the Brunswick "Fuze Eliminator." It features a "Low-Load Proactive® coverstock combined with a Low RG, High Differential core . . . While the core shape of the Brunswick Fuze Eliminator is the same as that of the Brunswick Fuze Igniter, the density profile within the core has been modified. By making the inner core heavier, the Overall RG has been lowered and the Differential RG raised compared to the Fuze Igniter. This produces a strong, heavy ball roll that allows the characteristics of the cover to show through." (And you were just going to pick up the ball sitting in the rack. . . .)

While a heavier, hooking ball certainly helps, don't discount the role of the pins it's attempting to knock down. Today's pins not only are lighter but also coated in a plastic that causes them to bounce around more. Since most pins are knocked down by other pins, that translates to more strikes.

But the biggest factor is the lane. Lanes have always been oiled to keep the balls rolling and prevent damage to the wood, but bowling alley proprietors have learned that oils can also be applied in such a way to steer the ball toward the head pin. The effect is subtle, and invisible to the naked eye, but has a significant effect on scoring. It won't pull your ball out of the gutter, but if you're generally around the headpin, it will help.

Indeed, the effect is significant enough that the American Bowling Congress has initiated an anti-oiling campaign, with the creation of what it calls "sport leagues." For these leagues, bowling alleys now have rules for where they can apply oil. The result has been a significant drop in scores, but an increase in the number of sport leagues, as better bowlers long for the "sport" designation and the opportunity for their skills to separate them from the herd.

Improving the Odds

To sum up, here's what you can do to improve your odds of bowling a perfect game:

1. Avoid "sport leagues" like the plague.
2. Buy a fancy ball and learn how to use it.
3. Practice a lot. Sadly, while a bad golfer can occasionally get a hole in one, a bad bowler will never roll a perfect game. You need to be rolling the ball in almost exactly the same place with almost exactly the same spin every time.
4. Send off to the American Bowling Congress for its *Bowler's Guide: An Instructional and Educational Book on Tenpins.* You'll learn not only the mechanics of the proper stance, approach, release, and follow-through but also some higher order strategies—you'll learn warm-up exercises, how to use positive mental imagery to raise your scores, and that "the number #1 enemy when bowling" is moisture in your shoes.
5. Don't start drinking beer until you've already had your first open frame. And *never* spill the beer in your shoes.

Becoming a Professional Athlete

The basic odds of becoming a professional athlete are not hard to figure. In the United States, the Bureau of Labor Statistics reported that in 2000 there were 9,920 professional athletes. With a total adult population of 221 million, that puts the odds at around 22,000 to 1.

If those odds don't discourage you, consider this fact: the mean annual wage of those athletes was $62,960. Given that those results were doubtless skewed upward by some very high salaries, the average "professional" is hardly making a killing.

So let's assume you don't want to join the Arena Football League or the North American Indoor Lacrosse League or play for the Durham Bulls or play golf in South Africa. You don't want to ride buses to events and sew your own uniform. No, what you want is the big-time money of the big-time sports; the private planes; the four illegitimate kids with three illegitimate women in two faraway cities; the call-in radio show where you mutter banalities like "We've got to make some plays if we're gonna turn this around" and "Hopefully, we'll get back on track" and they still give you a few bucks and a steak dinner for showing up. You want it *all*.

Well, then the odds shrink mightily, because you're really only talking about the major sports, and at least at this point you're talking only about men's sports. (Women in team sports make no money; the only ones who really cash in are the one-in-a-million tennis player or pixie ice skater.)

So, how many openings are there in the big-money leagues? Here's how the rosters look for the United States:

NFL	1,984 players
NBA	406 players
Major league soccer	276 players
PGA Tour	135 players
NHL	720 players
ATP men's tennis	1,345 players

Thankfully, the NCAA, doubtless concerned about professional sugarplums dancing in the heads of college athletes, has taken considerable care to calculate those odds for the major U.S. sports. They are presented below:

Estimated Professional Career Probability

Level	Men's Basketball	Women's Basketball	Football	Baseball	Men's Ice Hockey	Men's Soccer
High School to NCAA	2.9%	3.1%	5.8%	5.6%	12.9%	5.7%
NCAA to Professional	1.3%	1.0%	2.0%	10.5%	4.1%	1.9%
High School to Professional	0.03%	0.02%	0.09%	0.50%	0.40%	0.08%

Source: NCAA Research Staff; National Federation of State High School Associations

The NCAA numbers are a generous view of the odds in that they equate becoming a professional with being drafted by a professional team. In fact, many drafted players never make a professional team. In any event, roughly speaking, if you're a high school basketball player, your odds of being drafted by the NBA are roughly 3,300 to 1. The odds are only slightly better for football and soccer, and slightly worse for the WNBA. Baseball and ice hockey have significantly better odds, roughly 200 to 1 for baseball and 250 to 1 for ice hockey. In both baseball and hockey, though, there are extensive minor leagues where most draftees are sent; thus, the chances of a draftee actually making a Major League or NHL roster are far smaller than for an NBA, WNBA, or NFL draftee.

As hopeless as those odds may look, anyone considering a professional career must also consider another factor: length of career after making the professional ranks. For every household name, there are many players who only manage to hang on for a year or two, and then end up tending bar or selling Jeeps.

Average career length is actually a hotly debated statistical issue, particularly in the NFL, which given the violence of the game has the most career-ending injuries. The players contend with some justification that while fans may object to their six- or seven-figure salaries, the average player doesn't earn that money for very long, and doesn't exactly have a medical or investment banking career to fall back on.

In 2002, the NFL Players Association released a study of career length, using rosters from the 1987 to 1996 seasons. The total sample averaged 1,647 players a year, for a total of 16,000 player years. Its study showed the average career of an NFL player was 3.3 years. While the study did not speculate on the reason for the short careers, NFLPA officials said that the short careers are due primarily to the large number of high-speed collisions players experience. Consistent with that theory, the players who are hit the least, quarterbacks and punters/kickers, had the longest life spans—4.44 years and 4.87 years. The shortest careers belonged to those who were hit the hardest, running backs (2.57 years), followed by wide receivers (2.81) and cornerbacks (2.94). QED.

The journal *Chance* notes that a separate study conducted by a statistics class using 1970 data concluded that the average career length was 5.33 years, with even running backs lasting 4.35 years, and quarterbacks and kickers having around for 6.96 and 8.33 years respectively. Better news, but still not exactly life tenure.

Improving the Odds

You can't do much here to improve the odds beyond practicing your chosen sport and living clean, but you're best advised to try to pick up a degree just in case.

PART 5

MEDICAL MATTERS

Getting Hemorrhoids

Before we get down to the nitty-gritty, let's make you a few bucks. Bet a smart friend that you can name a word that he or she has used thousands of times, is in common usage, but that he or she cannot spell correctly. "Hemorrhoids" will win for you every time. Why is that? Well, first, it's not spelled the way it sounds, and the second "r" and adjacent second "h" are both silent. But the real reason that "hemorrhoids" is a difficult spell is that we avert our eyes whenever we see the word in print, because the thought is so unpleasant. (Kind of like "Camilla Parker Bowles.") And so begins our story.

The Odds

Estimates of the odds of getting hemorrhoids vary. According to Dr. Scott Thornton, a professor at Yale Medical School, there are 10 million hemorrhoid sufferers in the United States alone, leaving the odds of your currently having the condition at about 25 to 1.* The odds begin going up in middle age, with the prime hemorrhoid years being forty-five to sixty-five. The Mayo Clinic estimates that the odds are 1 to 1 over the course of a lifetime, while the manufacturers of the

* For those looking to probe this issue further, consider Dr. Thornton's helpful analysis in the *emedicine Journal*, available online.

Buttpillow (*www.buttpillow.com,* of course) estimate self-servingly that the odds are between 1 to 1 and 3 to 2 in favor. (Rather than using the Buttpillow at your office, you could simply hang a big sign around your neck saying, "I am a hemorrhoid sufferer; mock me.")

According to the U.S. National Digestive Diseases Information Clearinghouse, hemorrhoids is a condition where the veins around the anus or lower rectum become swollen or inflamed (sometimes referred to as varicose). The symptoms of hemorrhoids are frequently mistaken for other maladies. These symptoms, though, are similar to those presented by anal fissures and perianal dermatitis. Of course, you wouldn't exactly breathe a sigh of relief if your doctor told you that what you thought was a hemorrhoid was actually an anal fissure.

Not all hemorrhoids are created equal. In fact, we all have hemorrhoids; it's just that unless they produce symptoms, we don't notice them. But boy do we notice them when they do! Symptomatic hemorrhoids are of two types: internal (the really bad ones) and external (the really, really bad ones).

Internal hemorrhoids are given one of four grades. In increasing level of discomfort, they are:

- grade I hemorrhoids, which bleed but produce no other symptoms
- grade II hemorrhoids, which "prolapse"—that is, move outside the sphincter—but clear up on their own
- grade III hemorrhoids, which prolapse, and require you to push them back in with your hand
- grade IV hemorrhoids, which are permanently prolapsed

The diagnosis is not going to be a fun affair. For men, you can bet they'll be palpating the old prostate while they've got your legs spread. And then there's the "anoscope," which really needs no explanation.

Improving the Odds

How can you improve your odds? (For once, the author is convinced that he now has the reader's full attention.) Here are some important steps you can take:

1. *Stop reading on the toilet*. (Self-defeating advice from a book that was written for the toilet, but we're talking hemorrhoids here, and there is a moral imperative that transcends sales.) Dr. Thornton puts it best, "Prolonged sitting on a toilet (e.g., while reading) is believed to cause a relative venous return problem in the perianal area (a tourniquet effect), resulting in enlarged hemorrhoids."[11] When it comes to the old poop chute, you really want to avoid the "tourniquet effect."
2. *Be careful if you're pregnant*. The fact that pregnant women must suffer hemorrhoids in addition to all the other indignities of pregnancy is potent evidence against the existence of a benevolent god, but it is nonetheless a fact of life. You'll probably still decide to have children—if hemorrhoids are a deal killer, you probably weren't that enthusiastic anyway—but you need to stay off the toilet and heed the other advice below in order to reduce your odds.
3. *Eat a high-fiber diet*. Here's some counterintuitive news: your sphincter has to strain harder to produce small stools than it does to produce large ones. Since a high-fiber diet helps out in that department, you'll want to be eating your oats and grains.
4. *Don't squeeze too tight*. Again, tension in the rectal area is a problem. If nothing is happening, go back to whatever you were doing. Don't force it.

Even if these steps fail to prevent a hemorrhoidal onset, one good piece of news about hemorrhoids is that treatment is generally nonsurgical and highly effective. Simply correcting diet, toilet, and squeezing habits, as described above, resolves many cases. If those steps fail, the most popular treatment by U.S. doctors is known as "rubber band ligation." Yes, it's what it sounds like. You probably would have bet that there was no human malady that could be treated with a rubber band, but there you are.

If the rubber bands don't work, other nonsurgical treatments include carbon dioxide freezing, injections, and laser ablation. While these appear distasteful, they are all preferable to the most popular treatment of the Middle Ages: quite literally sticking a hot poker up the patient's ass.

Being Killed by Your Doctor

When the United States went from a military draft to an all-volunteer army, some criticized the move because it represented a further fragmentation of society. Where, they asked, do people of all races and social classes meet and spend time together? Here's the answer. Two places: the Department of Motor Vehicles, and the emergency room of a hospital. In those two places, we are all the same in the eyes of the authorities. And that's why we are all so afraid.

The Odds

According to a groundbreaking study by the National Academy of Sciences' Institute of Medicine in 1999, preventable medical errors kill between forty-four thousand and ninety-eight thousand Americans in hospitals each year. (Lest you think they plucked these numbers out of thin air, they are based on a Harvard study that examined thirty thousand patients at fifty-one randomly selected hospitals in New York, and a similar study of fifteen thousand patients at hospitals in Colorado and Utah.) That number makes preventable medical errors the eighth leading cause of death in the United States, ahead of automobile accidents and breast cancer.

According to other studies cited by the Institute, medication errors alone kill seven thousand Americans per year, more than workplace injuries (only six thousand dead). A follow-up study of thirty-six hospitals and nursing homes, reported in the *Archives of Internal Medicine*, found that one in five doses of medication was either the wrong drug or the wrong dose, or given at the wrong time or to the wrong patient. Another study conducted at two teaching hospitals found that almost 2 percent of patients were prescribed the wrong medicine. And these were *good* hospitals. Imagine the odds when the medicine is being prescribed in a *bad* hospital—or not even a hospital at all, but at a clinic or the office of a general practitioner.

Now, you might expect such a study to provoke considerable outrage, and it did. It did not, however, provoke considerable *action*. A follow-up report by the *Washington Post* three years later concluded:

> There's a lot of talk, but no significant progress. The reasons, observers say, include fierce resistance by doctors and hospitals to mandatory reporting and other Institute of Medicine recommendations, a lack of oversight by the federal government and the absence of an effective consumer lobby. . . .
>
> The nation's most exhausted and inexperienced doctors—the 100,000 interns and residents who staff teaching hospitals—continue to work as many as 130 hours a week, often with little or no supervision. Hospital-acquired infections, which kill about 90,000 patients annually, have increased 36 percent since 1980, a rise that coincides with the proliferation of bacteria capable of resisting the most potent antibiotics, according to the Centers for Disease Control and Prevention (CDC).[12]

After a lawsuit, hospitals have since limited interns and residents to eighty hours per week, but in a Catch-22, they are no longer allowed to sleep during that time, as they could previously. This message was brought home when many hospitals removed the bunks provided for that purpose.

As Geena Davis said in the 1987 remake of *The Fly* (another case of medical science gone wrong), "Be afraid, be very afraid."

Improving the Odds

Now that you know that there's a good chance your doctor may kill you, here's a problem: it's difficult to improve your odds.* The doctor who kills you probably isn't going to be the internist you see for your annual checkup. It's the doctor at the emergency room who ends up with your litter after some accident or seizure. And there you have limited control. Here are some tips:

- Say a little prayer that the date is not July 1 or shortly thereafter. July 1 is the date that residencies begin, and it is not a good day to be entering a hospital in a prone position.
- Keep in your wallet a card that includes your medical history, the medicines you take, and, most important, any allergies you have. If you have heart trouble, also include your electrocardiogram (EKG or ECG). You may wish to wrap the card in a $100 bill, just to get the nurse's attention. (Kidding!)
- Make sure you *understand* your medical history. For example, if you've had heart surgery, you need to know what kind, why, and when. Recognize that you need to explain your history in a way that the doctors will find *interesting,* and therefore remember in a pinch. (Ideally, your survival should not depend on your skills as a raconteur, but it does.)
- It never hurts to mention the words "chest pain." These words will move you rather quickly to the front of the line.

* While it contains no suggestions for patients, the Institute for Medicine's report does contain a helpful chapter on "Protecting Voluntary Error Reporting Systems from Legal Discovery." Now, that's a relief.

Best, though, to confine this strategy to cases where you have a legitimate belief that you are on death's door, or when you actually do have chest pain.

- Speaking of chest pain, you probably want to head to the emergency room when you first experience it, rather than finishing that last set of tennis or can of beer. (There is one activity, though, where you may persist, as death in this fashion has a cool name, "La Grande Mort," and a distinguished history.*)
- Ask questions. Sleepy interns need prodding to think. Also, recognize that you'll be answering the same questions from the nurse, the medical student, the intern, the resident, and the attending physician. It's called chain of command, and you need to deal with it cheerfully.
- Be very kind to your nurses. They are likely the ones who will spot errors. Don't be a pest, but when you have the chance, ask them about your course of treatment.
- Make sure you have a friend, relative, or even coworker accompany you to the hospital to act as your advocate. Don't be shy—buy them a nice gift later. Recognize that you're going to a dangerous place, and that you're not likely to be thinking clearly.
- Ask about an intensivist. "Intensivist" is a new and difficult-to-pronounce medical specialty that focuses on identifying trouble with intensive care patients. An intensivist monitors your chart to spot infections, pneumonia, or other problems that frequently kill intensive care patients. Studies show that

* Lord Palmerston, a nineteenth century British prime minister, died while having sex with a chambermaid on a billiard table. Former New York Governor and Vice President Nelson Rockefeller is best remembered for his grande mort atop his twenty-something mistress. French President Felix Faure's mort came in 1899. Pope Leo VIII experienced la grande mort with his mistress in 939; Pope Paul II with a page boy in 1471. In 453, Attila the Hun refused an emergency room visit and died on his wedding night, albeit not from a heart attack but a nosebleed.

your odds of dying in an intensive care unit are lessened by 30 percent if an intensivist is on duty. Better yet, over 20 percent of hospitals currently have intensivists working at least one shift, 10 percent have them working full-time, and the numbers are growing.

Dying of the Plague

"Plague." It has such a medieval ring to it. You can practically hear the trumpets in the distance and the rumble and creaking of the body-filled carts going by. But plague is hardly a thing of the past. What are your odds of expiring from a grisly "back to the future" plague? Unfortunately, higher than you might think.

Every age seems to have its plague, or pandemic. The most famous and destructive of them all, the Black Death of 1348–49, is estimated to have killed roughly 20 million people in Western Europe, or *a third of the overall population.* According to Norman Cantor's feel-good extravaganza *In the Wake of the Plague: The Black Death and the World It Made,* the Black Death was probably a combination of two diseases—bubonic plague (traditionally identified as the culprit) but also anthrax—spreading simultaneously. Bubonic plague is carried by fleas that live on the backs of rats (which gives the rats some deniability but doesn't make them any less revolting), but modern zoologists and biologists now believe that the Black Death spread in places (rural areas) and times (winter) where rats were unlikely to have carried the disease. Because anthrax and bubonic plague have similar flulike symptoms, and because cattle herds at the time showed evidence of disease, experts now believe anthrax was a factor as well.

The Black Death was exceeded in absolute, though not relative, scale by the Spanish flu outbreak of 1918, which killed 50 million people. While World War I was mourned as the senseless slaughter of a generation—10 million dead between 1914 and 1918—the flu outbreak of 1918 killed far more people, many of them soldiers in the cold, damp trenches of Northern France.

While these are the best known pandemics, there have been many others. Bubonic plague killed millions in the Byzantine Empire in the sixth century, and in southern Europe in the third century. Scientists now believe that a major cause of the decline and fall of the Roman Empire was an intense labor shortage caused by pandemics (first smallpox, then gonorrhea, then bubonic plague). Smallpox killed as many as 10 percent of the children in some countries in the seventeenth century; it also killed Queen Mary II of England, King Luis I of Spain, Tsar Peter II of Russia, and King Louis XV of France.

The American Indian population has been among the hardest hit people in the history of disease. The Indians of the New World had never been exposed to the diseases of the Old, so when European explorers landed, the results were disastrous. (The scientific term is descriptive; the Indian population was "immunologically naïve.") For example, the Spanish explorer De Soto was an evil man whose troops murdered, raped, and pillaged their way across what is now Florida and the American South. But his most murderous act was unintentional: bringing three hundred pigs along for the ride. According to researcher Charles Mann, a century after De Soto had passed through, the American Indian population had shrunk by *96 percent*, due mostly to smallpox carried by the pigs.[13] Other explorers spread their own diseases. Among the Incans, according to Mann, "Smallpox was only the first epidemic. Typhus (probably) in 1546, influenza and smallpox together in 1558, smallpox again in 1589, diphtheria in 1614, measles in 1618—all ravaged the remains of Incan culture."

The Odds

So in modern times, are our odds of experiencing a pandemic any lower? Today, we face perhaps the deadliest and longest-lasting plague of all—AIDS—the stirrings of another—mad cow disease—and the reminder, through SARS, that in the jet age, diseases can spread across continents very quickly. AIDS has killed 24 million people since it initially emerged in the early 1980s, and there are estimates that it will kill 70 million more over the next twenty. And future plagues of the types seen by our ancestors are inevitable.

Indeed, the parallels among latter- and modern-day pandemics are startling. Vegetarians will note with some justification that eating tainted meat has been the origin of most pandemics. Anthrax and mad cow disease come from tainted meat. Conventional scientific wisdom now is that AIDS originated in the "bushmeat trade"—that is the hunting and eating of chimpanzees—in the Central African rainforest in the 1930s.[14] Even more remarkably, one recent study, by Stephen J. O'Brien in the *American Journal of Human Genetics,* concludes that there is a genetic relationship between the Black Death and modern-day AIDS, and that if one of your forebears contracted and survived the Black Death six hundred years ago, you may now be immune to AIDS. In fact, 15 percent of the Caucasian population may be so immune.[15]

So, what are the odds? Well, if you assume that every generation (each thirty-year population group) suffers a plague that kills about 20 million people, then your odds of dying of a plague are approximately 200 to 1.

Improving the Odds

Going forward, our defenses against pandemic are far better than those of our Middle Age forebears. Better diagnosis and treatment, primarily through antibiotics, make for far better odds than in the time of leeches and hot pokers. Unfortunately, though, diseases adapt as well, as mutated strains become antibiotic-resistant, most recently in response to AIDS drugs.

We also face a significant threat of bioterrorism. Of course, this threat is hardly new. According to the U.S. Army Medical Research Institute of Infectious Diseases, biological warfare began in the fourteenth century, during the siege of Kaffa (now Feodossia, Ukraine) by a Tatar army. The Tatars catapulted into the city the plague-diseased corpses of some of their unfortunate comrades. And it worked! The plague spread, and the Tatars won Kaffa.

Lest you think that this sort of thing is only for (literal) barbarians, it turns out that the British Army was using biological warfare as early as 1753, in the French and Indian War. After an outbreak of smallpox among British troops, Sir Jeffrey Amherst, the commander of all British forces in North America, decided to "donate" blankets from the hospital to the Indian tribes opposed to the British, apparently as a peace offering. Once again, the immunologically naïve Indians succumbed.

But with many diseases like smallpox eradicated a generation ago, we have now grown "immunologically naïve" ourselves. So, sad to say, your odds of succumbing to pandemic probably are not significantly different from those of your ancestors of five hundred years ago.

Beating Cancer

Cancer. The big "C." Hearing that someone has cancer is among life's worst experiences. But such news is inevitably followed by two questions, "What kind?" and "Is that one of the bad ones?" We want to know the odds.

The Odds

Using data from the National Cancer Institute at the National Institutes of Health, here is a detailed look at those odds. As the graph on the following page shows, the odds for a cancer patient run from 9 to 1 against (pancreatic and liver cancer) to 9 to 1 in favor (thyroid and testicular cancer).

Improving the Odds

Thankfully, those odds are improving. While some forms, like pancreatic cancer, are an almost certain death sentence, the odds are now with you if you have breast cancer or prostate cancer. Sadly, though, survival rates vary for reasons other than cancer type. The 40 million Americans who have no health insurance have a poor chance at early detection, aggressive treatment, and eventual remission. Largely for that reason, survival rates for blacks are 10 to 15 percentage points lower than survival rates for whites. While it's amusing that your odds of dating a supermodel depend on how much money you have, it's not a bit amusing that your odds of surviving cancer depend on how much money you have.

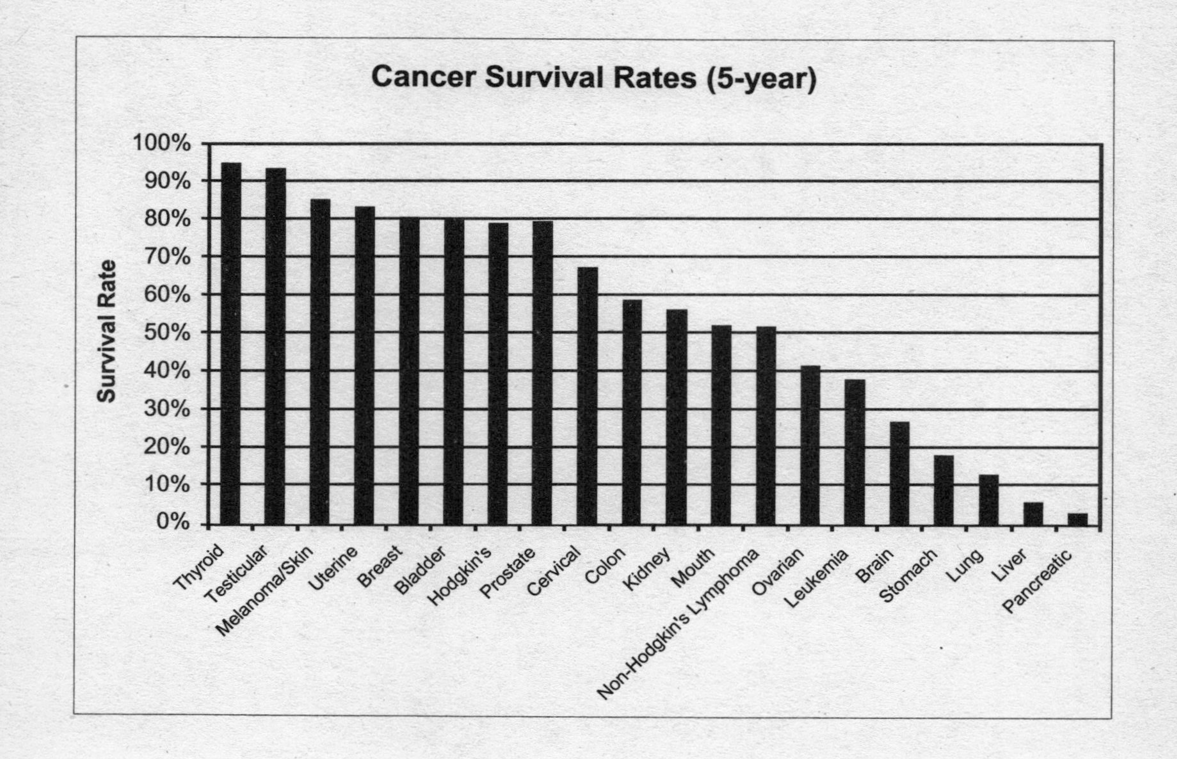
Cancer Survival Rates (5-year)
Survival Rate
100%
90%
80%
70%
60%
50%
40%
30%
20%
10%
0%
Thyroid
Testicular
Melanoma/Skin
Uterine
Breast
Bladder
Hodgkin's
Prostate
Cervical
Colon
Kidney
Mouth
Non-Hodgkin's Lymphoma
Ovarian
Leukemia
Brain
Stomach
Lung
Liver
Pancreatic

PART 6

THE HUMAN CONDITION

Being Six Degrees of Separation from Kevin Bacon (or Anyone Else)

You probably are well acquainted with the phrase "six degrees of separation" or its synonym, "the small-world phenomenon." It connotes that if you pick anyone else in the world, then the odds are that you are only six degrees of separation away from that person—that is, if you contacted someone you thought had the best chance of knowing your target, and that person contacted the person he or she thought had the best chance, and so on, then there is a better than even chance of your message reaching your target recipient in only six steps (that is, with five intermediaries).

You may not know, though, that the small-world phenomenon really originated with the psychologist Stanley Milgram. If you took Psych 101 in college, you've doubtless heard of Milgram. He became famous—actually, infamous—for the experiments he did on the willingness of subjects (volunteer college students) to impose painful electric shocks to other people whom they thought were other volunteers (actually, actors) in a study of the effectiveness of negative reinforcement. Basically, it turned out that American college students had a lot in common with Nazis in the "blame the victim" department. They were also emotionally scarred when they discovered as much.

But we digress. Stanley Milgram also conducted an experiment to determine how many acquaintances generally separate any two people chosen at random. Two earlier political scientists, Ithiel de Sola Pool and Manfred Kochen, had earlier theorized, based on a mathematical model that included population size and other factors, that the number was three, but in 1967 Milgram formulated an experiment to find out for real.

Milgram recruited volunteers in Kansas and Nebraska and told them the name, address, and occupation of a target person in Massachusetts. He then asked the Kansas and Nebraska "starters" to mail a document to someone they knew on a personal basis and whom they believed was likely to know the target on a personal basis. The results were remarkable. Milgram reported that "chains varied from two to 10 intermediate acquaintances, with the median at five"—in other words, six steps from starter to target. Milgram remarked at the time, "It's a small world."

The "six degrees of separation" reported by Milgram became a part of the lexicon due to the award-winning play of the same name by John Guare, which described the true story of a young black man who convinced a wealthy New York couple that he was Sidney Poitier's son. The subsequent movie starred Donald Sutherland, Stockard Channing (who received an Oscar nomination), and Will Smith.

Of course the oddest application of Milgram's small-world phenomenon is the game (and movie and websites, etc.) "Six Degrees of Kevin Bacon." The game was originated in 1994 by three Albright College students on a road trip, who talked their way onto *The Jon Stewart Show* (like *that* was hard!) and then produced a board game. For those who don't know the game, the challenge is to figure out the "Bacon number" for a given actor or actress—that is, how many movies removed from Kevin Bacon that person is. The late Orson Welles, for example, has a Bacon number of 2, having appeared in *The Muppet Movie* with actor Steve Martin, who in turn appeared in *Novocaine* with Kevin Bacon.

Interestingly, there is a high-brow, eggheaded version of Baconology. It involves degrees of separation from Paul Erdős. He was a Hungarian mathematician who published over fifteen hundred papers with 507 different coauthors. He is revered by fellow mathematicians, who feel ennobled if they were fortunate enough to be one of his coauthors—that is, to have an Erdős number of 1. Failing that, having published with someone who published with Erdős—an Erdős number of 2—is next best. There is (predictably) an Erdős Number Project on the Internet, at *www.oakland.edu/~grossman/erdoshp.html*.

Finally, the Milgram experiment has fostered an entire "small world" science devoted to exploring how people communicate across networks. This science is generally closely connected to computer science—think for example of the "six degrees of separation" as applied to how a computer virus spreads via email.

But just when we all started having fun with the small-world phenomenon and six degrees of this and that, along comes party pooper researcher Judith Kleinfeld of the University of Alaska at Fairbanks. A former student of Milgram's, she decided to do some follow-up research on the small-world phenomenon, went to the Milgram archive at Yale University to review his notes—*and discovered he basically made the whole thing up!*[16]

Milgram suppressed the results of a first study, where sixty subjects recruited through a newspaper article were assigned to reach the wife of a divinity student in Boston. Only three of the sixty starters (5 percent) managed to reach the target, and those three did so on average with nine degrees of separation. Moreover, in both this and the second reported study, Milgram did not select the starters randomly: in the first study, he recruited Wichita residents through a newspaper advertisement seeking people proud of their social skills and able to reach others across class barriers. In the subsequent reported study, starters in Nebraska were recruited from names obtained by purchasing mailing lists, tending to yield a higher income sample. Even then, only 42 of the 160 letters reached their targets.

Kleinfeld also found in the Milgram papers an unpublished study conducted by two other researchers, Beck and Cadamagnani, in 1968, and apparently sent to Milgram for review. When they replicated the Milgram study, Beck and Cadamagnani found that only 18 percent of the 150 letters they sent reached the targets. More significantly, they recruited as subjects fifty low-income people, fifty middle-income people, and fifty high-income people, and gave them a target of corresponding income. *None* of the low-income people were able to reach their targets.

These results suggest not that we all live in one small world, but that we live in very different worlds depending on our income, and that lower income people are far more isolated than those with money.

Think of it this way. If you're Alan Greenspan or Bono or Bill Gates, you're probably four degrees of separation from anyone on earth. Chinese peasant? Well, you know the Chinese Premier (or if there's been a recent coup and you aren't up to date, at least the American ambassador), who knows the head of the Party for the regional province, who knows the alderman (or whatever the Communist equivalent is), who, voilà, knows the peasant. But if you don't know the Premier or the ambassador—if in fact you don't know anyone outside your neighborhood, having spent your whole life there—then, no such luck.

Baconology, ironically, appears to be more valid than Milgramology. Go to *www.cs.virginia.edu/oracle/* and you will find a program run by the computer science department at the University of Virginia that will calculate the Bacon number for any given actor or actress. The average "Bacon number," according to mathematician Duncan Watts, is 2.918, and no actor on the Internet Movie Database is more than ten degrees of separation from Bacon. But Kevin Bacon does not have the lowest average score, in terms of being a hub for "connectedness": according to the the University of Virginia website, the Oracle of Bacon, Kevin Bacon only ranks 668th for connectedness among actors. (The top fifteen includes Robert Mitchum, Gene Hackman, Donald Sutherland, Rod Steiger, Shelley Winters, and Burgess Meredith.)

Sharing a Birthday

Having the same birthday as someone else is to most people a remarkable happening, and the best possible start to a relationship. For the clumsy male undergraduate, managing to produce the same birthday as a lovely coed is the best possible news, better than sharing a major or a dormitory. In many cases, it is sufficient to share a birthday with the coed's brother or father (though mother doesn't really work so well).

Warning to young men: reading the above, you may be tempted to prevaricate a bit in this department. Please remember that the true date of your birthday is no farther away than the license in your wallet. Concentrate on this fact really hard now, as it may be lost in the course of inebriation, yielding a colloquy like the following:

Male: *You know, Patty, we have the same birthday. Isn't that cool?*

Female: *My name is Cathy, and I've never told you my birthday. Let me see your license.*

Male: *Doh!*

The reverence for shared birthdays does not diminish as we get older and more sober. Fed chairman Alan Greenspan lunches every year on his birthday with former CIA and FBI director William Webster, former house speaker Tom Foley, and Senator Kit Bond. Why? Because it's their birthday (March 6), too.

Indeed, our love of shared birthdays not only does not wane as we grow older, but actually waxes. Followers of astrology aside, while for most of their adult lifetimes, most folks regard having the same birthday as another person remarkable, they regard having a birthday even one day apart of no consequence. Around about the time we turn sixty, and certainly by the time of the birth of any grandchildren, our tolerance for remarkability grows. A birthday within the same week—in some cases, within the same month, or even adjacent months—is cause for a telephone call. "You won't believe this, Ed, but my grandson's birthday is March 17 and *my* birthday is March 5 and my son-in-law's birthday is April 17—can you *believe* it?"

The Odds

So what are the true odds of sharing a birthday with the next person you meet? The odds are not 364 to 1, as you may believe, but rather vary quite significantly with your own birthday. That's because birthdays are not randomly distributed over the calendar, as the following chart shows:

Number of Births per Month

Month of birth	1997	1998	1999	2000	2001	5-year avg.	Percentage
January	317,211	319,340	319,182	330,108	335,198	324,208	8.2%
February	291,541	298,711	297,568	317,377	303,534	301,746	7.6%
March	321,212	329,436	332,939	340,553	338,684	332,565	8.4%
April	314,230	319,758	316,889	317,180	323,613	318,334	8.0%
May	330,331	330,519	328,526	341,207	344,017	334,920	8.4%
June	321,867	327,091	332,201	341,206	331,085	330,690	8.3%
July	346,506	348,651	349,812	348,975	351,047	348,998	8.8%
August	339,122	344,736	351,371	360,080	361,802	351,422	8.8%
September	333,600	343,384	349,409	347,609	342,564	343,313	8.6%
October	328,657	332,790	332,980	343,921	344,074	336,484	8.5%
November	307,282	313,241	315,289	333,811	323,746	318,674	8.0%
December	329,335	333,896	333,251	336,787	326,569	331,968	8.4%
Total	3,880,894	3,941,553	3,959,417	4,058,814	4,025,933	3,973,322	

Source: National Vital Statistics Reports, 1998–2002.

Thus, as we see, July and August are the most popular months, with February (by far), April, and November the least popular. February's unpopularity, though, is largely explained by the fact that it only has twenty-eight days.*

Not only are births not randomly distributed by month, they are also not randomly distributed by days of the week. Tuesday births are almost half again as likely as Sunday births. Of course, the day-of-the-week disparity is largely attributable to caesarean and other scheduled deliveries, which ob-gyns do not willingly perform on a weekend. Not only would they quite reasonably prefer to spend the weekends with their families, but also hospitals tend to be more leanly staffed.

That said, scheduled deliveries do not appear to fully explain the differences among days of the week. The same patterns hold true even for vaginal deliveries, which presumably are more likely to be unscheduled. Certainly, inducements using Pitocin and similar contraction-producing drugs—which now account for 20 percent of all births—tend to be scheduled for weekdays as well. But it does appear that even with an unscheduled birth, there is some likelihood that pregnant women, either consciously or subconsciously, can successfully direct their bodies to "hold it" until everyone reports for duty at the hospital.**

* Researcher Roy Murphy went further and used a database of 480,040 insurance policy applications to determine how birthdays were distributed by days of the month. He found that the first and fifteenth day of a month (e.g. June 1 or October 15) were significantly more popular than random chance would suggest, with the tenth and twentieth also above average in reported births. His theory, though, is that older persons had forgotten their birthdays and simply chose these obvious numbers to provide to the insurance company.

** Approximately 92 percent of births occur in hospitals, while 8 percent are what we'll call "banjo births," attended by midwives, psychics, or other folks (and named for the background music that is frequently playing during the delivery). Presumably, the distribution of banjo births should be more even, as no hospital visit is anticipated, but no study has been performed.

So if you were born on a Tuesday in, say, 1961, then the odds of matching birthdays with someone in your class is significantly greater than if you were born on a Sunday.

Improving the Odds

No discussion of birthday probabilities would be complete without a mention of one of the most enduring and counterintuitive questions in statistics: How many people must you invite to a party in order for the odds of two guests sharing the same birthday to be even—that is, to have a 50 percent chance? Think it over for a second, and proceed to the next paragraph.

If you are like most people, you probably thought the answer to the problem was $365/2 = 182$ or 183 people, which makes for a very large party. But not to worry, you won't need to spend the week cooking and can even open up the bar, because the correct answer is . . . 23.

Twenty-three?! There are any number of explanations of why 23 is the right answer. Here's one that's fairly easy to understand. Reverse the problem and consider the odds of *not* sharing a birthday. If you invite one guest to your party, then the chances of that person *not* sharing your birthday are 364/365. Add a third person, and the chances of that person *not* sharing a birthday with you *or* your first guest are 363/365. Given that we multiply to find the chances of events occurring simultaneously,* the chances of neither the first nor the second guest sharing a birthday are $364/365 \times 363/365$. If $n =$ the number of people invited to the party, then the chances of the guests not sharing a birthday are $364/365 \times 363/365 \ldots (365-n)/365$.

* So the chances of getting heads twice in a row are $1/2 \times 1/2 = 1/4$.

Thus, one way to solve the birthday problem is to determine the smallest possible number of guests (n) for which the above equation equals 1/2 or less—"less" since we're actually looking to match birthdays, as opposed to avoiding a match. In the event, when n = 22 guests (including yourself), the result is 0.52, but when n = 23 guests, then the result is 0.49. Now we're odds-on favorites.

If you still find this result counterintuitive, then note that the answer 183 is the right answer to a different question: how many people do you have to invite to have a 50 percent chance of one of them having the same birthday as *you*—that is, the same birthday as one person. That means each guest comparing his or her birthday only to you. If each guest is also comparing birthdays with every other guest, then the chance of a match rises quickly with each invited guest.

There's also a fun way to play this game out in real life if you have kids. Given that class size is generally somewhere around twenty-three, that means that two kids in your kid's class—though not necessarily your kid—should share a birthday every other year. You can do your own longitudinal study! (Plus, you get to *say* "longitudinal study," which makes you sound real smart. For bonus points, ask rhetorically, "Why aren't there any *latitudinal* studies?")

Being Overweight—and Losing Those Unwanted Pounds (For Good)!

In this chapter, we'll talk about weight, weight loss, and dieting—your odds of being overweight, and your odds of losing weight. Because this chapter must therefore go toe to toe with the thousands of weight loss books, videos, and seminars pressed on the public each year—all loaded with hyperbole and grand promises—it will require extraordinary means to capture and hold the attention of you, the diet-minded reader. Thus, while this chapter will adhere to the same rigorous scientific and mathematical standards used throughout the book, each topic sentence will conclude with an exclamation point.

The Odds

Let's start with the odds of your being overweight in the first place! A Harris poll commissioned in March 2002 concluded that 80 percent of Americans over the age of twenty-five are overweight, based on the Body Mass Index.[17] The Body Mass Index is calculated by:

1. multiplying your weight (in pounds) by 704.5
2. multiplying your height (in inches) by your height (in inches)
3. dividing the first result by the second

If you don't have a calculator handy, the National Heart, Lung, and Blood Institute at the National Institutes of Health provides a handy table at *www.nhlbi.nih.gov/guidelines/obesity/bmi_tbl.htm*.

According to the BMI, a score of 25 or more means you're overweight! So if you are six feet tall and weigh 184 pounds or more, then you are overweight. If you are 5' 6" tall and weigh 155 pounds, the message is the same. Of course, this means that many professional athletes and others we consider healthy are technically overweight.

Obviously, this topic is a hotly contested one: both altruistic nutritionists and crassly capitalistic weight loss marketers have an incentive to inflate the numbers in order to make us more receptive to their messages! No one really has much incentive to deflate (as it were) the numbers.

So let's say you've decided to shed some of those pounds! What are your odds? Studies show that if you go on a diet and succeed in losing weight, your odds of successfully keeping *any* of that weight off are 10 to 1. Of course, while these odds are bad news for you, the dieter, they are wonderful news for the multibillion-dollar weight loss industry, which counts on your fixating on the one in ten who is successful, concluding the problem was not your own willpower or physiology but rather the wrong diet, and trying again.

There is a good reason most diets fail, of course! The body tends to counteract dietary extremes—whether they be extreme intakes of food (all-you-can-eat resort) or extreme declines in food intake (desert island, Scotland). Conventional scientific wisdom now is that your weight generally will vary comfortably only within 10 percent of your natural weight. So, in other words, if you are a man who typically weighs 190 pounds, it will be difficult for you to maintain a weight of less than 181 pounds or more than 209 pounds.

If you are a 120-pound woman most of the time, then your expected range would be 108 to 132 pounds. For this reason, the National Institutes of Health recommends that the goal for any diet be weight loss of 8 percent over a period of three to twelve months, with the next goal simply being to maintain the new weight.

So what happens if you go beyond those extremes?! Three researchers recently presented a fascinating study.[18] They took one hundred volunteers whose weight had been stable for at least six months—that is, volunteers who appeared to be at their biologically "natural" weight. The researchers had them gorge themselves until they weighed 10 percent more than normal, and then had them diet until they lost the additional weight and went on to weigh 10 percent *less* than normal.

The study had some interesting results! At their normal weights, the subjects burned 1,360 calories per square meter of body surface per day. At 10 percent more than normal body weight, though, their metabolisms sped up and they burned 15 percent more calories than normal; conversely, when their weights were 10 percent below normal, they burned a mirror-image 15 percent fewer calories than normal. Basically, hormones act on the brain to tell the body to speed up or slow down, whether we want it to or not.

Don't get the impression, though, that your weight is in hormonal hands, and all attempts to lose weight are fruitless! One fact makes clear that behavior can have a significant effect on weight: over the past twenty years, the odds of your being obese have doubled. Incidence of type-2 diabetes, which is generally linked to obesity, increased by one-third in the 1990s alone. Because there is no reason to believe that our hormones are any more of an obstacle to weight loss now than they were twenty years ago, a change in behavior clearly has caused this communal weight gain. Indeed, the American Heart Association has specifically attributed the weight increase to the introduction of low-fat and nonfat foods; these foods, which are high in calories, lead us to eat more under the mistaken assumption that we won't gain weight.

A more intriguing theory blames Ray Kroc, the founder of McDonald's! No, not because his food is so fattening—that would only make McDonald's customers fat, and this is a broader problem. No, Ray Kroc invented "super-sizing," realizing that while consumers feel guilty and excessive going back for a second order of fries, they feel no guilt or excess in buying one "super-size" order of fries that's twice as large as the original. Thus, through a ripple effect (as it were), Kroc begat the 7-Eleven "Big Gulp" and the twenty-ounce plastic Cokes that now fill every grocery store aisle, and have driven their adorable eight-ounce glass progenitors into oblivion.

Improving the Odds

Now some quick advice on how to diet, and keep the weight off! Unfortunately, the advice is tragically boring, in line with the disheartening news above.

- *Exercise.* Not only does exercise help you lose weight, but also there is new evidence that even morbidly obese people can stay reasonably healthy if they exercise. So either way you're a winner.
- *Avoid empty calories.* It's bad enough that a can of Coke contains 140 calories, but those are empty calories. A recent

study shows that you're much better off consuming 450 calories of just about any food, even jelly beans, as food diminishes your appetite going forward. But drink a soda, and you just keep right on eating.

- *Set reasonable goals.* Again, according to NIH, 8 percent weight loss is reasonable, perhaps a little more if you're currently above your "normal" weight.
- *Think long term.* Cut out foods and super-sizes that you can afford to live without for the long term. There's no point giving up something you know you'll be eating again once you reach your goal, as you won't stay there for long.
- *Read more.* There's actually no evidence that reading causes you to lose weight, but it does help to feed authors, who as a group are tragically undernourished.[19]

Being Alone in the Universe

Talk about your enduring questions. For thousands of years, human beings have wondered whether other life is out there amidst the stars. That question has helped to spur advances in astronomy, physics, and numerous related scientific disciplines. It was the driving force behind *Star Trek, Star Trek: The Next Generation, Star Trek: Deep Space Nine, Star Trek: Voyager, Enterprise,* an innumerable series of *Star Trek* conventions, and the unfathomable fame and wealth of one William Shatner. Any idea powerful enough to make a millionaire out of a ham actor who speaks in staccato sentences must be a powerful idea indeed.

For most of human history, the question "Are we alone?" has really had no answers at all. While theories abounded, there was no real evidence either way. (No, crop circles and the Egyptian pyramids are not evidence.) Recent scientific advances, though, have changed things markedly. While we don't yet know the answer to the question "Are we alone?" we at least have a loose grip on how to calculate the odds.

To understand this better, we chatted with Gibor Basri, who is a professor of astronomy at the University of California at Berkeley and a fellow at NASA's Ames Research Center. Dr. Basri is a really smart guy. He says—in normal, everyday conversation—things like "What is the expanding universe expanding into? It is expanding into the *future.*" On the other hand, if you ask kindly, Dr. Basri will also explain cosmology and astrobiology to you on a fourth-grade level. We have found our guide to the odds of intelligent life in the universe.

The Odds

To explore the question, Dr. Basri begins with what is known as the Drake Equation, named for Dr. Frank Drake, who originated it in 1961.* The Drake Equation is the following: $N = R^* \times fp \times ne \times fl \times fi \times fc \times L$. Don't panic and skip to the next chapter, though, for Dr. Basri will break it down into bite-sized pieces. But before we move on, realize what the Drake Equation is trying to calculate (N). It is not the odds that there has ever been intelligent life in the universe, or even the odds that there is currently life in the universe, but rather the odds that we could currently *find* intelligent life in the universe, or in Drake-speak, the number of civilizations in the Milky Way Galaxy whose electromagnetic emissions are detectable.

An easy way to think of the Drake Equation is this: the number of planets with intelligent life can be determined by multiplying the rate at which such civilizations appear by the length of time they exist. (If civilizations appear frequently but last only a millennium or two, they will be hard to detect; so, too, will civilizations that are enduring but appear only rarely.)

According to Dr. Basri, determining the rate at which intelligent civilizations appear involves six factors, set forth below.

A star is born. First, for life, we need a star (let's just trust him on that one), and we thus begin by determining how often stars appear in the galaxy. Given that there are already about 300 billion stars in a galaxy that is 10 billion years old, the rate would appear to be thirty per year. But given that the rate of star formation has slowed in recent millennia, Dr. Basri thinks twenty per year is a fairer estimate.

* You can learn more about the Drake equation at *www.seti.org*. "SETI" is an abbreviation for Search for Extraterrestrial Intelligence. The fancy acronym tends to generate more credibility than, say, *www.littlegreenmen.com*.

Planet. Next, we need a planet circling a star, which will be the incubator for life (as the star itself is too hot and vaporous). So the next factor is the probability that a star will have planets. Fortunately, the formation of planets is a natural outgrowth of star formation. Using current techniques, astronomers are detecting a probability of 0.05 (about 20 to 1 odds). But given the limits of those techniques, it's certain that there is significant undercounting going on. According to Dr. Basri's estimates—and we're in no position to argue, now are we?—we're probably missing nine out of ten planets using our rudimentary methods, so the probability that a given star will have a planet is actually around 0.5 (even odds).

Habitable planet. The next question is the odds of these planets being habitable—that is, capable of sustaining some form of life. The planet can't be too close or too far, but must be in what astrobiologists call the "habitability zone." (New Yorkers will understand this as the astronomical equivalent of living within a block of the Park; residents of Charleston should think "south of Broad Street.") Of course, the best evidence that planets tend to be habitable is the fact that the earth is habitable. Indeed, life can be found in some of Earth's least habitable places—areas of extreme heat, areas of extreme cold, Cleveland. We thus assume that other planets like our Earth—same size, same temperature—are most likely to be habitable under the right conditions. And life could arise in other environments as well. Total it all up and the probability of habitability is about 0.1, or 9 to 1 odds.

Simple life. Next, we calculate the odds that a habitable planet actually ends up holding life—not necessarily intelligent life, but at least simple, single-celled life. (Note that this question differs from the previous one; just because a planet can sustain life does not mean that it will produce life; think of a habitable planet as a flower awaiting a bee.) The answer, though, appears to be that habitable planets almost always end up holding at least simple life. Consider that our own planet was hardly a hospitable home to life in its early days. At one point it was a gigantic snowball. At other points, it was consumed by poisonous gases. It has endured periods of violent meteorites. Not to mention the Disco Era. And yet single-cell life emerged very quickly. So let's assume that simple life on a habitable planet is a certainty—a probability of 1.0, no odds.

Pause here and note that given the math thus far, astronomers are fairly confident that there are and will continue to be millions, possibly even billions of planets in the Milky Way Galaxy with simple life on them. But that is not the end of the story.

Complex life. Simple life doesn't really make a planet that interesting to us. There weren't a lot of episodes where Captain Kirk or Captain Picard fought a pitched battle with a protozoa, or hooked up with a comely amoeba. From the perspective of the Drake equation, radio conversations with single-celled organisms tend to be decidedly one-sided (sort of like interviews with monosyllabic professional athletes). So the next step in the chain is to determine the probability that simple life will evolve into complex, intelligent life. Here, the probability is not so great. Again, we can only generalize from our own experience, but consider that it took the earth—so quick to establish simple life—over 3 billion years to move up to complex, intelligent life. So, this is not an easy trick. Let's drop the probability here to 0.1, or 9 to 1 odds.

Contact. Even with multicellular life, though, you do not have what astrobiologists call "technical civilizations"—that is, civilizations capable of communicating with us through radio transmissions or other sophisticated means. It took us earthlings quite a while to reach the point where we were both able at and interested in communicating with others in the galaxy; in fact, many people on our planet live in communities that have not reached this point at all. Thus, Dr. Basri's estimate—which he acknowledges is almost completely a matter of personal taste—is a probability of only 0.01, or one in a hundred that intelligent life will be able to, and choose to, communicate with others.

Time. At this point, if you multiply all of our probabilities together, then you have an equation of "average lifetime during which technical civilizations emit radio transmissions/1,000 years." What does that mean, though? That means that one thousand years is the minimum average lifetime needed to make it likely that there is at least one technical civilization around at all times. But in order to determine the number of intelligent civilizations in the galaxy, we need to multiply that number by the average length of time such civilizations exist. If such civilizations on average exist for a thousand years, then multiplication tells us that there should be only one such civilization in our galaxy—and thus, that we are reasonably likely to be alone. If we believe civilizations last for ten thousand years, then we should have nine companions.

Distance (and time, again). Of course, that doesn't necessarily mean that we're going to be able to find them, because we live in one mighty big galaxy. Our galaxy has a diameter of one hundred thousand light-years, so any current radio transmissions from the outer reaches of the galaxy are almost certain to reach our solar system long after the earth is gone or uninhabitable. Ditto for our own transmissions. And remember that conversation requires a round trip. If you assume that our nearest neighbor is "only" one thousand light-years away, then the minimum lifetime required rises to the order of 10 million years.

If you assume that our civilization (and others like it) will last 10,000 years, then the odds are that there will be ten other technical civilizations transmitting at the same time, and that we will hear from them. In other words, the odds are that we are not alone, and will confirm as much in the next few millennia.

Improving the Odds

There's nothing we can do to improve the odds of intelligent life in the universe (other than preserving our own planet), but there are ways we can improve our odds of *detecting* any such life. One of NASA's near-term missions, scheduled for flight in the next five years, is the Kepler Mission. The Kepler Mission will put into orbit a one-meter telescope designed to detect habitable planets. It will do so by noticing the tiny variations in light that occur when a planet's orbit takes it between Earth and the star it is orbiting.

The Kepler Mission should profoundly expand our understanding of planets. Current techniques for planet finding are able to detect only huge, Jupiter-sized planets, and only sixty such planets have been found so far. But consider this: the Hubble Space Telescope was able to view a swath of the sky the size of a grain of sand held at arm's length, and then only for about a half hour at one time. The Kepler telescope will view a swath of sky the size of your hand held at arm's length, and do so continuously and with better resolution. In this way, the Kepler telescope will observe about one hundred thousand stars in the Milky Way galaxy for at least four years. Thus, Kepler should be able to find those Earth-sized planets with the best chance of sustaining intelligent life. For more details, check out *www.kepler.arc.nasa.gov*.

And stay tuned. We should know for sure in the next thousand years or so.

GIVING BIRTH TO A GIRL

Nowhere do the odds appear to be more regularly defied than when it comes to the sex of one's children. How often do we meet a couple who, though initially inclined toward a boy, has produced two girls; they decide to keep trying, and produce a third girl, and perhaps a fourth. Is this to be expected? What are the odds?

The Odds

You might expect the odds to be even, but they are not. The odds of having a boy are consistently better than the odds of having a girl, even if only marginally. In 2000, there were 2,076,969 male births and 1,981,84 female births, making the odds against having a girl 1.048 to 1. (In other words, there are 1,048 males born per 1,000 females.) The longer odds of having a girl have proven persistent, as we see over the past fifteen years:

Annual U.S. Births By Sex

Year	Girls	Girls %	Boys	Boys %
2001	1,968,011	48.9%	2,057,922	51.1%
2000	1,981,845	48.8%	2,076,969	51.2%
1999	1,932,563	48.8%	2,026,854	51.2%
1998	1,925,348	48.8%	2,016,205	51.2%
1997	1,895,298	48.8%	1,985,596	51.2%
1996	1,901,014	48.9%	1,990,480	51.1%
1995	1,930,234	49.2%	1,996,355	50.8%
1994	1,930,178	48.8%	2,022,589	51.2%
1993	1,951,379	48.8%	2,048,861	51.2%
1992	1,982,917	48.8%	2,082,097	51.2%
1991	2,009,389	48.9%	2,101,518	51.1%
1990	2,028,717	48.8%	2,129,495	51.2%
1989	1,971,468	48.8%	2,069,490	51.2%
1988	1,907,086	48.8%	2,002,424	51.2%
1987	1,858,241	48.8%	1,951,153	51.2%

Source: National Vital Statistics Reports/Centers for Disease Control

The same is true over longer periods, as the odds have varied by less than 1 percent over the past sixty years. The odds, though, have moved ever so slightly in the female direction, as the odds against having a girl were 1.055 to 1 in 1940.

Why are boys born more frequently than girls? No one knows for sure, but some speculate that it is an evolutionary adaptation. Because boys and men tend to die more frequently than girls and women at every given age, they need a "head start" at birth, if there are to be enough males left over to propagate the species. (Interestingly, the imbalance is even greater at conception, with more than half again as many male embryos as female, but female embryos survive in vitro far better than male embryos.)

Interestingly, the ratio varies across ethnic groups. Japanese mothers face the longest odds against having a daughter (1.084 to 1), while black mothers face closer to even odds (1.031 to 1).

Improving the Odds

In 1970, Dr. Landrum Shettles published a book, *How to Choose the Sex of Your Baby*, which as the title would suggest had some easy-to-implement strategies for sex selection. Most of the techniques stemmed from Dr. Shettles's conclusion that sperm carrying Y (male) chromosomes are less adept at surviving in an acidic environment than sperm carrying X (female) chromosomes. For a bunch of yucky reasons, this acidity theory argues for couples seeking a girl to: (1) refrain from having sex around the time of ovulation, (2) have sex in the missionary position, and (3) have the woman avoid orgasm. Reverse the advice if you wish to have a boy.

Dr. Shettles's methods have ancient origins. According to the Talmud, if the woman's orgasm precedes the man's, any resulting child will be a boy. (The Talmud does not address the case of the man having an orgasm, going off to watch the Jets game, and the woman having no orgasm at all.)

Hundreds of thousands of people swear by Dr. Shettles's advice. On the other hand, since he has sold over a million books, the odds would tell you there should be half a million people swearing by his advice even if it is total poppycock. Some researchers, though, have concluded that the Shettles method has the opposite of its intended effect, and others believe it reduces the odds of conception. A peer-reviewed study in the *New England Journal of Medicine* concluded that is has no effect whatsoever.[20] Or, in *New England Journal of Medicine* talk, "There was no evident relation between the age of sperm and the viability of conceptus." (You probably thought "Conceptus" was the Roman general that Russell Crowe played in Gladiator.)

For those truly adamant about sex selection and willing to

spend thousands of dollars, however, there is new, sounder technology. An organization called the Genetics and IVF Institute offers couples a method of improving the odds dramatically. The MicroSort technology is based on the fact that an X chromosome (which in sperm produces a girl) is substantially larger than a boy-producing Y chromosome. Since chromosomes are made of DNA, human sperm cells having an X chromosome will contain approximately 2.8 percent more total DNA than sperm cells having a Y chromosome. Through some fancy scientific techniques they call "flow cytometric sperm separation technology"—undoubtedly to disguise the fact that they're really just using a teeny tiny sperm strainer—MicroSort takes sperm, separates the X from the Y, and artificially inseminates the mother-to-be with the chosen type.

Although the system is hardly foolproof, the odds are affected dramatically: according to MicroSort, of couples seeking females, 91 percent have been successful; of couples seeking males, 76 percent have been successful.* Lest you doubt this is on the up and up, the MicroSort technology is licensed under a patent held by the U.S. Department of Agriculture and is in clinical trial with the FDA. Plus, it's been featured on the CBS *Early Show*!

So, if you're Mrs. Baldwin and you've already produced Alec *Hunt for Red October* Baldwin and Daniel *Homicide: Life on the Street* Baldwin, you can stop the madness. You don't need a Billy *A Pyromaniac's Love Story* Baldwin or a Stephen *Celebrity Mole: Hawaii/Bio-Dome/Slap Shot 2* Baldwin. You can turn to producing actresses.

* Of course, some see serious ethical problems with enabling sex selection. MicroSort, though, does offer discounts to couples attempting to avoid sex-linked diseases, such as hemophilia.

MULTIPLE MOUTHS TO FEED: THE ODDS OF TWINS (OR MORE)

Talk about your late nights and panicky days. The odds of having two (or more) children of the same age are definitely worth considering. While there is an upside—you can place them in a TV sitcom, as child labor laws require two kids to play the role of one—you'll be shelling out a lot of dough for two of everything. So what are the odds? The answer: not too long, and getting shorter.

Number of Multiple Births (and Percentage of Total Births)

Year	Total	Twins		Triplets		Quadruplets		Quintuplets or more	
		Actual	Odds	Actual	Odds	Actual	Odds	Actual	Odds
2000	4,058,814	118,916	34 to 1	6,742	602 to 1	506	8,021 to 1	77	52,711 to 1
1995	3,926,589	96,736	40 to 1	4,551	862 to 1	365	10,758 to 1	57	68,888 to 1
1990	4,158,212	93,865	44 to 1	2,830	1,469 to 1	185	22,477 to 1	13	319,862 to 1

Source: CDC National Vital Statistics Report; *www.twinstuff.com*

Of course, the significant shortening in the odds of multiple births is largely or wholly attributable to the increased use of fertility drugs. One study of triplet and higher number births concluded that 43 percent resulted from the use of assisted reproductive techniques such as in vitro fertilization; 38 percent resulted from the use of ovulation-inducing drugs; and only 20 percent were spontaneous. Assuming those percentages hold true for twins, the odds of having twins without taking fertility drugs lengthen to around 170 to 1.

PART
7

VIOLENCE AND TRICKERY

Getting Away with Murder

You thought O.J. was the only one to manage it? Think again. Your odds may be better than you think. Clearance rates for homicides have been in a steady decline since the 1960s and, as we'll see, the odds are to some extent in your hands.

The Odds

According to FBI crime statistics, the nationwide clearance rate for murder is 62 percent—odds of about 2 to 1 against your succeeding in your murder. Your odds are actually better than clearance rates would indicate, though, since the FBI counts a case as cleared once an arrest is made, even if the case is subsequently dismissed or the defendant found not guilty. (Clearance rates also generally tend to underestimate your odds of success because many crimes—most notably, rape—go unreported; this factor is not significant with murder, however, because, well, there is a body lying around, and people tend to notice. Similarly, when multiple offenses are committed, only the most serious crime is reported, but since murder is trump, it does not suffer underreporting on this count either.)

Improving the Odds

One important influence on your odds of success is where you commit the crime. Since all major cities use the FBI's

Uniform Case Reporting System to report clearance rates, it is fairly easy to pick your most advantageous murder locale. Your best odds are in medium-sized cities (those with five hundred thousand to 1 million residents) where the odds of your getting away with murder are about even (54 percent clearance rate). Apparently, such cities offer the benefit of urban anonymity but do not have as professional a police force as larger cities. The one place to clearly avoid is a rural county, where the odds against your escaping are 4 to 1 (a 77.5 percent clearance rate). In these Mayberry RFD–type places, everyone knows everyone else's business, and out-of-town murdering types tend to get noticed.

An organization called the Justice Research and Statistics Association (*www.jrsa.org*) has done a helpful study on other factors affecting whether the police can solve a murder, from which we've distilled these helpful hints:

- *Avoid witnesses.* In the JRSA study, almost half of all cases were closed because witnesses at the scene identified the murderer.
- *Remember to leave the scene of the crime.* It's surprising, given that this advice has been in the public domain since the cop shows of the 1950s, but many murderers forget this simple fact. In the JRSA study, the police listed "Offender arrested at or near the scene" as the reason for closing 18 percent of all cases.
- *Don't make a habit of this.* In over 10 percent of cases, the police found their murderer by linking the method of the crime to him or her. Here's a tip: if your nickname is "Stiletto Steve" or "Three-in-the-Chest Theo" or "Arsenic Anna," you probably need to stop the murdering, find a new modus operandi, or find a new town.

The factors affecting closure rates cited in the JRSA study tend to focus on the identity of the murderer and victim, and how the crime was investigated. Other studies, though, show some other important things to keep in mind:

- *Do not confess*. The *Washington Post* did a detailed study of police practices in Prince George's County, Maryland.[21] The *Post* found that 68 percent of cases were closed by a confession. While confession rates have varied significantly over time and place, it's safe to say that at least 25 percent of all cases require a confession to be closed. (Confession rates actually are studied quite frequently, generally by those who support overturning the Supreme Court's 1966 *Miranda* v. *Arizona* ruling, whose holding you have probably heard ten thousand times on television: "You have the right to remain silent. You have the right to an attorney . . ." Now, you'd think that having the police tell you to keep your mouth shut and to hire a lawyer would tend to cut down on confession rates. And the studies show that it did have that effect for property crimes such as burglary and robbery. But confession rates for murder and other violent crimes were unchanged after *Miranda*. While no study has rigorously examined why, the theory here is that violent people tend not to listen and are generally stupid.)

  Now, a clarification for less alert readers: confessing does not include just telling the police that you committed a crime. Leaving a "calling card" in your own handwriting saying "I am God" counts as confessing. Telling your cellmate "You know, I killed a guy last month" counts as confessing. Confessing to a radio deejay counts as confessing. If you're in a confessing mood, think about talking to a priest, and remember to speak softly, in the little booth provided for this purpose.

- *Do not murder any member of Jessica Fletcher's family, or anyone in Cabot Cove, Maine*. Her closure rate is 100 percent, and she is exceptionally annoying to boot.

Being Unjustly Convicted of a Crime

While some of us wonder whether we could commit a crime and get away with it, others fret that they might be convicted of a crime committed by someone else. Does this happen? (1) Yes. (2) But only if you're poor!

Reliable data on unjust convictions is impossible to obtain, of course, given that there's no way to study the issue systematically. We can, however, get a clue from what has happened with DNA testing, which works only when police are able and inclined to gather blood, hair, or similar human evidence. Such evidence is unusual, and most likely in serious cases of rape or murder.

According to the Innocence Project at the Cardozo School of Law, over 120 innocent people have been released from prison since DNA testing was commenced in 1989. Twelve innocent prisoners have been released from Death Row. Those numbers, though, are only the tip of the iceberg:

- Only 29 states allow convicted prisoners access to DNA testing.
- Most prisoners don't even know to ask.
- As noted, in most cases, there is no original or residual evidence on which to perform DNA tests.

So while there is no way to know for sure, you probably want to keep a little money in the bank for DNA testing and a good mouthpiece, just in case.

Murder—Being a Victim

Thanks to *America's Most Wanted* and similar shows, Americans are well aware that there are sinister people in our midst. Watching the local news, you'd think America's streets are rivers of blood. But what are the true odds of meeting a violent, involuntary death? A lot longer than you think.

The Odds

Each year in the United States, there are approximately sixteen thousand homicides, leaving the crudest measure of the odds against being murdered at 18,000 to 1.

Race and sex have a great deal to do with your odds of being murdered. Sixty percent of murder victims are black males, while only 9 percent are white males. Black females (10 percent) fare even worse than white males, and the safest group of all is white females (only 2 percent of all victims, despite being the largest population group).

Improving the Odds

Regardless of your race and sex, though, there is a lot you can do to lower your odds.

- *Stay away from open-air drug markets.* About 28 percent of all murders occur in these places.

- *Be careful whom you let into your car.* Fourteen percent of murders are committed in a car.
- *Don't join a gang or drug organization.* One quarter of all murder victims are so affiliated.
- *Avoid a life of crime.* Eleven percent of murder victims were committing a crime at the time of their demise.
- *Stay away from young males.* Half of all murders are committed by males ages fifteen to twenty-four.
- *Stay away from people with guns.* Over 65 percent of all murder victims are killed by a handgun; another 10 percent are killed by other guns.
- *Choose your friends wisely.* Over half of all victims know their murderer. So when you meet someone at work or a party and are considering whether to ask that person over for a drink or out to a ballgame, run through your checklist (Is he a young male? Does he want to ride in the car with me? Does he have a handgun? etc.)
- *Realize that names matter.* According to an item in *Chance News,* attorney William Linka conducted a study in 1997 of the names of persons convicted of capital murder and eligible for the death penalty. He used state legal records and opinions from the U.S. Supreme Court (where most death penalty appeals end up). His discovery was intriguing: 8.3 percent of the persons were named Lee; 5.4 percent were named Dwayne, Duane, or DeWayne; 19.4 percent had no middle name; 13 percent were a "Jr." or "Sr." And a total of 41 percent of those convicted for capital murder met at least one of those criteria. The predictive power of Mr. Linka's theory has recently been buttressed by the arrest of John Lee Malvo for alledgedly acting as the Washington, D.C., beltway sniper and Derrick Todd Lee as the alleged Louisiana serial killer. So, if David Lee Roth drives by and

asks you to hop in and see his new gun, even the most ardent Van Halen fan should think twice.

So Do I Buy That Gun or Not?

We're not going to get into a whole Second Amendment Charlton Heston v. Sarah Brady debate here. Just the odds. In other words, if you're considering the purchase of the gun, are the odds with you or against you—put another way, what are the odds of that gun killing you (or a member of your family) versus the odds of that gun killing a criminal attacking you (or a member of your family)?

From the National Safety Council, we know the odds that a gun in your home will end up killing you. The odds of a fatal, accidental discharge are 11,000 to 1. The odds that you will commit suicide with your handgun are 214 to 1. (Of course, some of these folks would resort to poison or tall bridges if there were not a handgun present, but a fair number of impetuous suicides would be prevented.) So those are the odds of the gun costing you your life.

Here, according to the FBI's Uniform Crime Statistics, are statistics on how often using that gun will *save* your life.

Justifiable Homicide
by Weapon, Private Citizen,* 1997–2001

Year	Total	Total fire-arms	Hand-guns	Rifles	Shot-guns	Fire-arms, type not stated	Knives or cutting instru-ments	Other danger-ous weapons	Personal weapons
1997	280	238	197	16	14	11	28	6	8
1998	196	170	150	6	14	0	17	5	4
1999	192	158	137	5	10	6	18	9	7
2000	164	138	123	4	7	4	15	8	3
2001	215	176	136	10	13	17	25	6	8

* The killing of a felon, during the commission of a felony, by a private citizen.

Now, ordinarily when you see a chart with itty bitty numbers like these, you look to the upper left-hand corner to see if those numbers are "in thousands" or "in millions." Well, you'll see no such notation here. So, in 2001, for example, a grand total of 176 Americans used a firearm to kill an attacker, odds of about 1.5 million to 1.* (Of course, there are certainly other cases where the brandishing of a firearm prevented the brandisher from death or bodily harm, but note that *the gun doesn't have to be loaded to be brandished.*) Not to be judgmental, but those odds sure look a lot larger than the accidental discharge and suicide odds.

* The great majority of these cases no doubt involved a domestic disturbance that would not otherwise have ended fatally, but let's assume that in each case, the firearm purchase was a lifesaver.

Having Your Identity Stolen

Identity theft can take many forms. The 1998 film *Face/Off* presented an extreme case, when an FBI agent (John Travolta) underwent extensive plastic surgery to replicate the appearance of a comatose terrorist (Nicolas Cage) in order to penetrate his network. Once revived, the terrorist Cage underwent equally painful surgery to take on the appearance of FBI agent Travolta. As pure cinema, it is hard to beat John Travolta pretending to be Nicolas Cage pretending to be an FBI agent when he is in fact a terrorist, while Nicolas Cage is pretending to be John Travolta pretending to be a terrorist when he is in fact an FBI agent. Outrageously, Oscar took no notice.

More recently, we saw Harry Potter and his friend Ron Weasley employ a polyjuice potion in order to take on the appearance of their nemeses Crabbe and Goyle. While there are rumors that author J. K. Rowling considered having Crabbe and Goyle do the converse, she no doubt feared accusations of a "*Face/Off* ripoff." But those concerns aside, identity theft will no doubt remain a cinematic staple. (Not to mention the soap operas. . . .)

For the average nonterrorist muggle, though, identity theft is likely to take a more mundane, but certainly troublesome form: a criminal impersonating you through the mails (thereby obviating the need for surgery or potions) in order to assume your identity for purposes of obtaining a credit card or other loan. What are the odds of this happening to you? Hold onto your hat.

The Odds

Identity theft is the fastest growing crime in the world. The Identity Theft Resource Center estimated that there were over seven hundred thousand cases of identity theft in 2001. The Federal Trade Commission, which sponsors a Consumer Sentinel website where consumers can file a complaint of identity theft, reports that complaints on its site alone grew from 31,117 in 2000 to 86,198 in 2001 to 161,189 in 2002. The FTC estimates that identity theft costs the financial services industry, which generally reimburses customers for losses, over $1 billion per year. (Of course, much of that amount is passed back to the consumer in the form of higher fees or interest rates.) Your odds of having your identity stolen in the coming year are probably around 200 to 1.

According to the FTC, the uses to which identity thieves put this information break down as follows:

Credit card fraud	42%
Phone or utilities fraud	22%
Bank fraud	17%
Employment fraud	9%
Government documents/benefits fraud	8%
Loan fraud	6%
Other	16%

(You have to wonder how often the identity thieves actually get away with utilities fraud, since the thief is presumably paying the utility bill at his own house, but go figure.)

The way identity thieves obtain your personal information has spawned numerous colorful names. "Shoulder surfing" involves watching or listening as you state or key in your telephone calling card number or credit card number over a public phone. "Dumpster diving" involves going through your garbage to obtain credit card or bank statements. More recently, "spam" email uses some pretext to ask the recipient for personal financial information.

The mother lode for any surfer, diver, or spammer is your social security number. That's because over time the SSN has grown to become a universal identifier—how everyone from your credit card company to your hospital identifies and tracks you. With your social security number and some other financial information in hand, a thief can access your existing accounts or, worse yet, establish *new* accounts, using a different address. That makes it harder for you to spot the theft, because you won't be able to notice unauthorized activity on your statements—you don't even know you have the account. Your first notice will come when those accounts are maxed out and overdue, the credit reporting bureaus are notified, and your existing credit lines are cut off.

There is some good news, however. First, for new accounts, since you never authorized them, you're not liable for any of the charges on those accounts. Second, for your own account, federal law caps your liability at $50 for a lost or stolen card; you have no liability for any charges incurred after you have notified the card issuer that the card is lost or stolen. To encourage on-line use of credit cards and allay concerns about identity theft, many issuers have voluntarily lowered that liability to zero.

Improving Your Odds

There are several simple steps you can take to reduce your risk of identity theft:

1. Don't feel like you need to spend a lot of money on this problem. A cottage industry has grown up to calm your fears of identity theft—for a price. Most of the work is being done by the nation's credit reporting agencies, which will be the first to know if your credit suddenly suffers the high volume increases in number of accounts and ensuing charges that attend identity theft. But the odds of suffering a major loss for which you are personally responsible are still sufficiently long that you probably don't want to be paying someone an annuity to avoid it.
2. Read your monthly statements promptly upon receipt. Check to make sure that there are no unauthorized charges. This will not be difficult, as identity theft is not a subtle crime, generally involving large-screen TVs and other electronic devices. If you see "Circuit City—$8,000" or "Best Buy—$6,000," then you may have a problem.
3. Stop your mail when you're traveling. You'd be surprised how much someone can learn about your finances by reading a week's worth of mail.
4. Beware of crooked relatives. As noted above, credit card issuers and banks are legally required to remove any authorized charges from your account. Unfortunately, this law has bred another type of crime. Some cardholders will arrange for a cousin or a friend to "steal" their account information, run up a bunch of charges, and then split the goods with the cardholder. The cardholder then notifies the bank that the card has been stolen, and gets all the money refunded. The perfect crime.

This crime, despite being inherently immoral, also has an unfortunate side effect for honest victims of identity theft. Your repayments may be held up while the credit card company investigates whether your information was genuinely stolen. If it turns out the thief was your brother-in-law or cousin, then you can expect that investigation to increase in intensity and lengthen in time.

5. Finally, try not to get too paranoid. Identity theft is a real problem, but not a reason to live in a cabin in the woods. In particular, it is no reason to retreat from the world of electronic commerce. Reputable websites encrypt your credit card account number, and do not ask for any additional personal information. While there have been cases where someone—generally, an employee—gained unauthorized access to all the account numbers at an on-line merchant, no loss to customers resulted. Just remember, you're more than happy giving your credit card to an unshaven waiter at an out-of-town bar; you shouldn't be any more nervous providing it to the good people at amazon.com.*

* The above example is provided with some hope that amazon.com will not, like all of its on-line cousins, fail spectacularly, thereby dating the book.

PART
8

THE COLOR OF MONEY

Avoiding an IRS Audit

Every type of taxpayer, regardless of culpability, worries about an IRS audit. People who are scrupulously honest tend to be worriers, so they worry. People who are less honest but have rationalized their misdeeds (think waiters, think tips) worry that they may have overdone it. People who are blatantly dishonest (think drug kingpins) may worry the least, since they have so much else to worry about.

The Odds

The truth is, everyone would worry less if they understood the odds. Audit rates for the IRS are at an all-time low. The odds of being audited in 2002 were 175 to 1, having fallen from 95 to 1 just since 1992. So, in other words, at 2002 rates, if you began paying taxes at age twenty-one and live to be seventy-two, the odds of your experiencing at least one audit during your lifetime are 3 to 1.

The odds are even better if your fear is showing up at an IRS district office for a face-to-face audit. You've had the nightmare, or seen it played out in the movies: there you are, with your shoebox full of receipts, suddenly remembering some income you forgot to report, or a deduction that may be dubious. Facing you is a man with thick, dark-rimmed glasses wearing a yellow button-down shirt with those unfortunate short sleeves and three ballpoint pens in his shirt pocket. Your underarms begin to dampen, and you wish that you, too, could wear short-sleeve, button-down shirts.

Well, if this is your concern, the odds are very much in your favor. Face-to-face audits at IRS district offices have plummeted. The IRS increasingly favors what it calls the "service center" audit—which is a fancy way of saying an audit conducted through the mail. Perspiration aside, though, the service center audit tends to generate almost as much revenue as the face-to-face audit. (One benefit for the taxpayer, though: with no IRS official to whom to lie, there is less chance of going to prison for lying to a government official.)

The reason for the decrease in audit rates and the shift to service center audits is not difficult to understand. Staffing at the IRS has been cut steadily over the past decade. Furthermore, temporary employees have come to represent a larger percentage of the Service.

IRS Employment

IRS Fiscal Year	Service Total	Permanent	Temporary
1992	114,819	113,028	1,791
1993	110,680	109,133	1,547
1994	109,505	108,215	1,290
1995	114,064	96,787	17,276
1996	102,082	91,832	10,250
1997	97,404	87,101	10,303
1998	97,375	84,317	13,058
1999	97,526	82,350	15,176
2000	97,464	82,023	15,441
2001	100,577	83,007	17,570

Source: TracIRS, Internal Revenue Service, Office of Personnel Management

Improving the Odds

Your chances of audit, though, depend greatly on your own particular circumstances. For example, you would logically assume that the IRS would audit wealthy people more frequently than poor people. After all, in terms of generating needed funds for the Treasury, a rich tax cheat is a much more attractive target than a poor tax cheat, as catching the rich yields a much higher return on the investigative dollar than catching a poor one. Furthermore, taxpayers whose only income is a salary have little room for cheating, as that salary is reported directly to the IRS on a W-2.

This sound logic actually used to apply, but throughout the 1990s, audit rates rose for the poor, and fell for the rich. The reason is largely political: when President Clinton sought to expand the earned income tax credit for the working poor, the Republican Congress agreed only on condition that the IRS be directed to audit more of those receiving the credit, forcing it to shift resources. The result now is that you have a better chance of getting an audit notice on an assembly line than at a yacht club. To wit:

Audit Rates and Results by Income

	Low income (<$25,000)		High income (> $100,000)	
	1992	2001	1992	2001
Audit rate (%)	0.4	0.78	2.38	0.31
Avg tax recommended	$2,471	$2,577	$4,569	$4,567
Avg audit hours	0.9	1.5	2.4	1.8
Number of audits	171,864	325,441	74,566	29,086

Source: Transactional Records Access Clearinghouse, Syracuse University

As the chart shows, in 1992 the rich were six times more likely than the poor to be audited; by 2001, the poor were more likely to be audited than the rich. Furthermore, and quite remarkably, whereas auditing the relatively simple returns of a low-income taxpayer had previously taken 37 percent of the time of it took to audit a high-income taxpayer, that number had risen to 83 percent ten years later.

Next, you might assume that your audit odds are the same no matter where you live. This is certainly not true. For example, in 2001, if you were a high-income taxpayer in Los Angeles, your audit odds were about 200 to 1—three times greater than if you lived in Georgia (600 to 1). Lest you think it's an urban/rural thing, consider that the odds in Arkansas and Oklahoma (135 to 1 or 0.74 percent) were more than three times as great as in New Jersey (435 to 1). The odds seem to vary pretty randomly, based simply on the personality of the folks at your local district office.

These data are really embarrassing—to the IRS and the government at large. So what would you expect the IRS to do in response? Audit the poor less? Audit the rich more? If you guessed either of these, you clearly live outside the Beltway. The correct answer is: "Stop gathering the data!" In 2001, the Bush Administration announced that it would no longer report individual or corporate audit returns by region. Problem solved.

There are other factors besides income and geography that will affect your odds of audit.

First, in order to decide whether you are an audit candidate, the IRS runs a computer program that compares your return to an "average" return, generating what is called a "DIF" score (likely short for "different"). So, for example, if the average person with your income claims $10,000 in deductions, and you claim $50,000, then your DIF score will rise. The same is true for charitable deductions (particularly if you forget to submit Form 8283 for in-kind donations over $500) and unreimbursed business expenses.

Second, if your occupation is one that involves handling a lot of cash—waiter, dentist, entertainer—then your odds of being audited increase.

Third, and perhaps most importantly, if you are self-employed and therefore filing Schedule C (Sole Proprietorships), then the IRS reasons that you have more opportunity to hide income or fabricate deductions, and your chances of audit triple. Claim the home office deduction, a frequent area of abuse, and the chances are still greater.

Lengthening your odds of audit is difficult. You should *not* give up deductions to which you are entitled, stop giving to charity, or trade in your cool small business and schedule C for the daily grind of corporate life and a less audit-provoking W-2. Of course, you should file an honest return, but that does not diminish your chances of having an audit; it just increases your chances of surviving one.

There are really only two actions you can take that will reduce your chances of audit without increasing your taxes or harming your quality of life. Both are advice you first heard in grade-school mathematics:

Check your work. Forgetting your Social Security number, making mathematical errors, and forgetting to report income that an employer has reported on a W-2—all these little mistakes can help to provoke an audit. This is less true now that the IRS's computers correct some of these mistakes automatically, but it never hurts to get it right.

Show your work. If an IRS agent is handed your file and is wondering why you claimed an unusually large deduction, a typewritten explanation and receipt attached to your return might easily allay any suspicions. Absent that explanation, the agent's only choices are to accept the return as filed or order an audit. Your best strategy for avoiding an audit is to make the agent's life easier.

Striking It Rich on *Antiques Roadshow*

The most relaxing hour in television must be PBS's *Antiques Roadshow*. Watch as a bunch of people empty their attics and present the contents to antiques dealers. Watch as some of them are delighted to find out that the dreary landscape that used to hang in Uncle Joe's study actually conceals an equally dreary—but far more valuable—Renaissance masterpiece.

How often do these things really happen? We spoke to Marsha Bemko, the senior producer of the *Roadshow*, and asked her to help figure out the odds.

The Odds

Antiques Roadshow has become a cultural phenomenon. The *Roadshow* now visits four U.S. cities each year, and televises three or four episodes from each city.* Once in residence, the *Roadshow*'s seventy to eighty experts—antiques dealers and appraisers primarily—review whatever the townsfolk happen to bring in.

The first obstacle to striking it rich on *Antiques Roadshow* is getting in the door. The show has become so popular

* The *Roadshow* visited thirteen cities in 1997, its debut year, staying only one day per city. It visited eight cities each year from 1998 to 2001, and seven in 2003. The shows producers then decided to make longer visits in fewer cities.

that you actually need a ticket for admission. Applicants send in a postcard, and the *Roadshow* holds a random drawing. On average, the *Roadshow* gets 83,000 applications for each stop, but currently is able to hand out only 10,000 of tickets. So the odds are 7 to 1 against even getting the china cabinet out of the pickup truck.

Once in the door, the odds remain long. With 10,000 tickets, that generally works out to only 5,000 households, as the winners tend to bring a spouse; however, each pair of contestants is permitted to bring two items, so that means that the *Roadshow* ends up evaluating around 10,000 items per stop, or 40,000 items per year. Over the nine-year life of the program, that's 360,000 items in all.

Each of these items is evaluated by an expert, who might consult other experts on-site, someone they know back at their home office, or Internet listings of auction results. If the expert likes what he or she sees, the expert then flags down Ms. Bemko to pitch her on why you and your treasure should appear on the *Roadshow*. Of course, all the experts like to appear on camera, and are eager to pitch their own "clients"; Ms. Bemko chooses among them, and sends the lucky guests and experts off to the green room for makeup and antiques stardom. Given that only about fifty items per visit are televised, the odds against even appearing on the show are around 200 to 1.

But what are the odds of striking it rich? Well, you're not going to become a millionaire: the most valuable item ever discovered on *Roadshow* was a Ute First-Phase chief's wearing blanket, which was brought to the Tucson stop in 2002. The Navajo blanket, which the owner had kept folded over the back of a rocking chair, was worth $350,000 to $500,000. If we consider a quarter of a million dollars striking it rich, there have been only a half dozen items of that value in the history of the show, so given the number of entries, the odds are 60,000 to 1. Drop your threshold for riches to $100,000, and the odds shorten to 10,000 to 1.

Improving the Odds

If you'd like to improve your odds on the *Roadshow*, you'll need to exercise some judgment about what you bring to the appraisal table. According to Ms. Bemko, your best odds are with furniture and paintings. Grandpa's old gun or the teakettle your great aunt brought back from Japan probably aren't going to cut the *Roadshow* mustard.

Interestingly, your odds are much better if you are British rather than American. *Antiques Roadshow* actually began in the United Kingdom in 1977—a tradition that has been carried on by shows like *Who Wants to Be a Millionaire* and *Changing Rooms* (and its U.S. equivalent, *Trading Spaces*).

The odds for UK contestants are better all around. The *Roadshow* makes many more stops, what with it being a small island and all. There is no need for tickets or a lottery; anyone can participate. And the odds of finding a valuable antique are far greater because the country is far older than its former colony.

"WINNING" AT KENO

One of the biggest differences between Las Vegas now and Las Vegas ten years ago is the ubiquity of keno. Keno boards litter the casinos. You can play keno on the television in your room. Keno runners will take your bets while you order dinner at a restaurant. And when you return home from Vegas, your local lottery probably offers a keno-based game.

Why is keno so popular with the casinos and lotteries? Because (1) the odds are fantastic for the house, and (2) the odds are impossible for the average inebriated casino patron to understand.

Keno is a game of chance, akin to bingo, and requires no skill to play. Eighty balls numbered 1 through 80 are put in a cage or bowl, and the casino draws out twenty of those balls, one at a time. You gamble on which numbers will be drawn by filling out, before the drawing, a keno card. The card has eighty numbered squares on it, and you cross out the ones you believe will be chosen. In most games, you can pick anywhere from one (a "Pick 1" game) to fifteen (a "Pick 15" game) numbers. Your payout in a given game depends on how many numbers you pick and how many you match. Generally, you need to match at least half of your chosen numbers to earn any payout; the more numbers that match, the more you earn. The greatest payout, obviously, comes from selecting a lot of numbers and matching all of them.

The Odds

For any casino game, the expected value of any given bet involves two variables: the chance that you will win the bet, and the odds paid by the casino if you win. (Note that in the gambling context, the "odds" do not refer to your chances of winning, but to what you're paid if you do win.) Thus, for example, your chance of winning a bet on black in roulette is 47.37 percent;* the odds offered by the casino on such a bet—the payoff—is 2 to 1. So the expected payoff is the chance of winning times the odds, or 94.74 percent. Since 100 percent is a fair bet—that is, one where over time you will end up winning back the money you wager, but no more—that means over time you lose at roulette.

The bottom line for casino games is referred to as the "house percentage" or the "hold." This number represents, over time, the amount of each bet the house can be expected to retain. So, for example, a house percentage of 10 percent means that, given your chances of winning and the expected payout, the house will end up keeping 10 percent of every bet you make—$10 on a $100 bet—whereas you end up keeping 90 percent, or $90. In our roulette example, the house advantage on a bet on black (or red) is 5.26 percent—the percentage of green space, where everyone loses, on the wheel.

Matters get complicated, though, because even within a particular game, the house percentage can vary. As an example, consider a sampling of the bets available in craps and the

* The odds are less than 50 percent because the wheel includes eighteen black compartments, eighteen red compartments, and two green compartments (numbered "0" and "00"). If you bet black, the house wins on red or green. If you bet red, the house wins on black or green.

typical payout. Even if you don't understand craps,* you can see that the house percentage varies significantly by how you choose to bet—as low as 0.58 percent, which is basically even money odds, and as high as 11.11 percent—almost twenty times poorer odds.

House Percentage at Craps (selected wagers)

Bet	Chances of Winning	Payout/ "Odds"	House Percentage
Pass line	49.30 %	1:1	1.41 %
Don't pass	50.71 %	1:1	1.36 %
Come	49.30 %	1:1	1.41 %
Come bet, 3 to 1 odds	49.30 %	1:1	0.61 %
Don't come	50.71 %	1:1	1.36 %
Don't come, 3 to 1 odds	50.71 %	1:1	0.58 %
Place Bet			
6 and 8	45.45 %	7:6	1.52 %
5 and 9	40.00 %	7:5	4.00 %
4 and 10	33.33 %	9:5	6.67 %
Hard Ways			
6 and 8	9.09 %	9:1	9.09 %
4 and 10	11.11 %	7:1	11.11 %

* Here it is in a nutshell. You or another player (known as the "shooter") rolls two dice. If the shooter rolls 7 or 11, then you win; if the shooter rolls 2, 3, or 12 (collectively, "craps"), you lose. Any other roll—say, a 6—and the shooter rolls again. The shooter continues rolling until he or she rolls either that same number (6) again—in which case you win—or a 7—in which case you lose. We say "win" here, but one of the unique factors of craps is that you can actually bet against the shooter (even if you are the shooter) as described below.

In craps, you can place bets both before and after the first roll of the dice. A "pass line" bet is a bet that the shooter makes 7 or 11 on the first roll, as described above; "don't pass" is the opposite. After the initial roll (assuming it's not a 7, 11, or craps), a "come" bet means a bet that the shooter will repeat his first roll before rolling a 7, as described above; "don't come" is the opposite. A "place" bet is effectively a side bet that the shooter will roll a chosen number before a 7. A "hard way" bet is similar, but on an unusual combination of two numbers—say two 2s to reach a 4—rather than a combined total. There are other bets not included in the chart.

As you can see from the chart, some of the bets in craps offer very hospitable odds, with a "come" or "don't come" bet offering 3 to 1 odds, meaning a house percentage of less than 1 percent. But there are some very bad bets as well, with some "place" and "hard way" bets offering worse odds than roulette.

With that perspective, let's consider what makes keno different from all the rest.

First, the house percentage is astronomical. It generally varies from 25 percent to 30 percent. Casinos have one defense, which is that keno is an expensive game to administer, as it takes up a lot of floor space and requires the employment of runners to pick up the tickets. But don't doubt for a second that keno is a major profit center. And if you want better odds, you're always free to pick a game that doesn't require floor space and runners.

Second, the house percentage varies from casino to casino, changes frequently, and is neither intuitively obvious nor publicly disclosed. Whereas the odds at roulette, say, are widely disseminated and constant across the industry, keno is a black box. Sure, you can *derive* the house percentage if you take a calculator to the casino—and some players do—but that's hard.

To get a firmer grip on how all this plays out in reality, we contacted Michael Shackleford, the "Wizard of Odds," who runs a website of the same name. The Wizard actually tracks down the odds by visiting casinos. To get an idea of how the house percentage works, and varies depending on your bet, consider two cards from the Tropicana Casino in Atlantic City, as analyzed by the Wizard.

Keno—Pick 3 Odds

Catches	Pays	Pick 3 Probability	Return
0	0	0.416504382	0
1	0	0.430866602	0
2	1	0.138753651	0.138754
3	43	0.013875365	0.596641
Total		1	0.735394

Keno—Pick 10 Odds

Catches	Pays	Pick 10 Probability	Return
0	0	0.045791	0
1	0	0.179571	0
2	0	0.295257	0
3	0	0.267402	0
4	0	0.147319	0
5	1	0.051428	0.051428
6	22	0.011479	0.252547
7	150	0.001611	0.241671
8	1000	0.000135	0.135419
9	5000	6.12E-06	0.030603
10	100000	1.12E-07	0.011221
Total		1	0.72289

Source: The Wizard of Odds

The two cards track the total expected return (the converse of the house's percentage or hold) if you were to pick three or ten keno numbers. The total return is derived by adding up the expected return for each possible outcome—for the Pick 3, that means "catching" zero, one, two, or three of your chosen numbers. As you'll see, if you pick three numbers, then you win nothing in the event either none or one of the numbers shows up. If two numbers show up, then your bet pays 1 to 1. The probability of that outcome is .139. If all three numbers show up among the chosen twenty, then your bet pays 43 to 1 odds. But the probability is only .014. The total expected return on the card is the sum of the expected return for each possible outcome (0, 1, 2, or 3 catches). Here, your expected return is 73.54 percent, and the house hold is therefore 26.46 percent. For the Pick 10, though, your expected return drops to 72.29 percent, and the house hold rises to 27.61 percent. So, in other words, your expected return is higher (the house hold lower) with the Pick 3 than the Pick 10. But given that all the casino presents to you is the first two columns (the payout per number of catches), you could be forgiven for not understanding that nuance.

How much do the odds vary from casino to casino? The Wizards of Odds surveyed the house percentage at every Vegas Casino for one type of keno bet: the Pick 9. Those returns varied tremendously—from 20.15 percent (by far the best) at Silverton, to 28.13 percent at Bellagio, to 29.77 percent at Circus Circus, to 44.74 percent (the worst) at Monte Carlo. Of course, with keno, those percentages have doubtless changed many times since, and not necessarily for the better.

Improving the Odds

Let's be clear. With some considerable effort, you can improve your odds at keno. Improving the odds requires you to go from casino to casino, discovering the ticket price and payout rate for each bet, matching those numbers with the probabilities, and determining the casino's hold. You can certainly perform this calculation with a pocket calculator. The formula is relatively simple, and the probabilities are available in chart form from the Wizard of Odds and any number of keno books and newsletters.

But the odds will never approach what you can get at other casino games, because the odds always stink. Thus, it is hard to know how people can sell books with names like *Winning at Keno!* or *Keno Winning Strategies!* to the gambling public. Those titles are the moral equivalent of *Enjoying Your Herpes!* or *Skiing in Jamaica*. It is simply a contradiction in terms.

That said, there are a few select people who have proven consistent winners at keno. Donald Trump is a consistent winner. Steve Wynn is a consistent winner. And there are others. What do they have in common? *They own the casinos.* Keep it in mind.

Winning at Blackjack

Blackjack is unique among all casino games. Blackjack is a game where the skill of the player shapes the odds to a remarkable extent. It is also the only casino game where a skilled player can push those odds to even or better than even. *In blackjack, uniquely, the odds are up to you.*

The rules of blackjack are very simple. You are dealt two cards, as is the dealer,with one of the dealer's faceup and one face down. You can then choose to take additional cards (hit) or stand pat (stick). If you hit and go over 21, you "bust" and lose the game. If your final total is 21 or below, you win if your hand is closer to 21 than the dealer's.

The blackjack dealer plays his or her hand according to established rules, and exercises no judgment. Most house rules require the dealer to stick on 17, hit on 16, and follow a few other simple decision rules.

The standard hand pays 1 to 1 odds, though a hand of 21 is "blackjack" and pays 1.5 to 1.

The Odds

The true odds of winning at blackjack are difficult to calculate for the average player, who bases his or her decisions on subjective feelings about the state of play. Those decisions might be based on how much money the player has won or lost, whether an attractive waitress or waiter is nearby, and whether he or she is "feeling lucky." All of these judgments are frequently affected by the ingestion of copious amounts of alcohol, thoughtfully provided by the casino. For these players, there are neither calculable odds nor meaningful hope.

Rather, our look at odds will focus on five levels of rational blackjack strategy. "Rational" here means having one or more objective decision rules, and following them consistently, in contrast to the "gut feel," "hot chick," or "Jim Beam" strategies. They vary according to how they approach (1) the player's own hand, (2) the dealer's hand, and (3) the amount of the wager.

Improving the Odds

The basic imitate-the-dealer strategy. Followed to some extent by most players, this strategy focuses primarily on the player's own hand and simply imitates the decision rule of the dealer: hit on 16 and stick on 17 and above. What the dealer is holding is noted only in extreme cases (e.g., if the dealer is showing a 10, face card, or ace). The simplicity of the strategy leaves plenty of time for drinking and talking, and is followed by many a player.

It is interesting to consider why this strategy is an exceptionally poor one, which it is. After all, shouldn't following the same strategy as the dealer result in even odds? The problem is that, because you must play your hand before the dealer, you go bust first. Put another way, if the dealer sees you go bust, the dealer doesn't play out his or her hand and call it even if the dealer busts, too; the dealer just takes your money.

For that reason, the follow-the-dealer strategy gives the house a 5.5 percent edge over you. That's still better than some other games, like keno, but you can do better.

The one-bet, dealer-observant strategy. The greatest boost to your odds of winning at blackjack comes if you take account of the dealer's hand in shaping your own strategy. Thus, in deciding whether to hit or stick, you are considering two factors: the card(s) in your hand, and the dealer's face card. The decision rules to implement this strategy can be memorized in advance, which should only take a few hours. Numerous books and websites offer the strategy in chart form, with some suggesting that it is some sort of proprietary "system." Basically, it's just the answer to a math problem. Generally, there are two charts to memorize, one for "soft" hands (where there is an ace that can be valued as either 11 or 1) and one for "hard" hands (no ace, or ace counted as 1 because 11 would bust you). Here's what a chart for a hard hand would look like:

Your Hand	Dealer's Up Card									
	2	3	4	5	6	7	8	9	10	A
11 or less	Hit	Hit	Hit	Hit	Hit	Hit	Hit	Hit	Hit	Hit
12	Hit	Hit	Stick	Stick	Stick	Hit	Hit	Hit	Hit	Hit
13	Stick	Stick	Stick	Stick	Stick	Stick	Hit	Hit	Hit	Hit
14	Stick	Stick	Stick	Stick	Stick	Stick	Hit	Hit	Hit	Hit
15	Stick	Stick	Stick	Stick	Stick	Stick	Hit	Hit	Hit	Hit
16	Stick	Stick	Stick	Stick	Stick	Stick	Hit	Hit	Hit	Hit
17–21	Stick	Stick	Stick	Stick	Stick	Stick	Stick	Stick	Stick	Stick

Conceptually, this strategy reflects both the odds of your going bust *and* the odds of the dealer going bust. Thus, for example, when you are holding cards worth 12 (say, a king and a 2), your odds of going bust on the next card are only about 35 percent; nonetheless, the chart advises you to stick in cases where the dealer is showing a 3, 4, or 5. Why? While your odds of going bust are low, the odds of the dealer going bust are sufficiently high that it is not worth taking that risk.

In reading this chart, recognize that if you wish to do well playing blackjack, you cannot treat it as a guideline or a suggestion. These are rules. It is *never* rational to do otherwise based on feel or instinct. If you follow these rules, then the advantage the house holds over you drops precipitously, to only around 1 percent. That's extraordinarily good, considering that your room only costs $99 and you just got a free breakfast.

The variable-bet, dealer-observant strategy. The advantage of the one-bet, dealer-observant strategy is that it improves your odds to close to even without requiring a lot of work. Moving those odds higher gets a bit more complicated. It involves taking advantage of some rules of blackjack that offer significant but only occasional opportunities to improve your odds. Those rules involve doubling down and splitting pairs.

- Doubling down allows you to double your bet and take one additional card. Given that the card most likely to be drawn next is a 10 or a face card worth 10, then you should almost always double down if your initial total is 11. But, depending on what the dealer's up card is, the odds may dictate doubling down on any total between 8 and 18.
- If you are dealt two cards of the same value (say, two 3s or a 10 and a king), then you may split them, double your bet, and play two hands. Absent card counting information, splitting aces and 8s is always the correct play; splitting 4s, 5s, and 10s is never the correct play; and the correct play on other pairs depends on the dealer's up card.

Deciding the correct times to double down and split pairs requires memorizing several additional charts akin to the one above. There are also variations on this strategy depending on whether the casino where you are playing uses a single deck or a multiple deck. But if you do that work, your odds rise to basically even money. Think about that: with a little memorization and some self-discipline, you can play all night and rationally expect to walk out with about the same amount of money you walked in with. In a casino. Wow.

Card counting. Card counting is based on one fundamental fact about blackjack. It is the only game in the casino with a *memory*. One turn of the roulette wheel does not depend on the previous turn; the numbers picked in one keno game do not affect the next. But in blackjack, when the cards are not shuffled after every hand, the odds change depending on what transpired on the previous hand.

If you have any doubt that blackjack is really a game about mathematics rather than instinct, consider that card counting was not developed by experienced players in the casinos of Las Vegas but rather made its first appearance in an article in the *Journal of the American Statistical Association*. Entitled "The Optimum Strategy in Blackjack,"* the article was published in 1957, and the ideas contained therein were quickly recognized and improved upon by Edward Thorp, then an associate professor of mathematics at the University of California at Irvine. His book *Beat the Dealer* became the bible of card counting.

Card counting strategies depend on one fundamental concept. High numbered cards favor the player, while low numbered cards favor the dealer. Since the dealer has no choice but to hit on totals of 16 or lower, the dealer must hit—and likely bust—*even if every remaining card in the deck is a 10.* Furthermore, given that blackjack on the initial deal

* Among the founding fathers of card counting were the authors, Roger Baldwin, Wilbert Cantey, Herbert Maisel, and James McDermott.

is a 1.5 to 1 payoff for the player, and given that a 10 and an ace are the most frequent blackjack combination, a proliferation of 10s also improves the player's odds of blackjack.

Contrary to popular conception, card counting does *not* mean memorizing every card played since the last shuttle. Rather, it means keeping a running total "score" based on the cards that have been played—in the most elemental version, assigning one value (+1) to the low cards (say, 2–6) and another value (-1) to the high cards (say, 10 and the face cards). As each hand is played, you assign a value to each card revealed, and add those values to your running total. You need only keep in mind that running total in deciding whether and how much to bet, as it represents how favorable the deck is to you. *Et voilà,* you're a card counter!

More refined strategies—valuing more of the cards, varying the values assigned—can improve the odds further. And of course you have to practice enough that you can quickly (and subtly) count all the cards, even if there are multiple players at the table. Done right, though, basic card counting can rather easily lower the house percentage to −2 percent.* What that means is that *you* have the advantage; the longer you play, the more likely you are to win.

MIT-level card counting strategies. While certainly improving the odds, traditional card counting systems have problems

* Given the potential for card counting to improve the player's odds, and the ability to defeat this strategy by simply shuffling the cards, you might wonder why casinos simply don't shuffle after each hand. The answer is simple: shuffling costs them money. Every second the dealer spends shuffling is a second that he or she could be dealing. And given that only a tiny minority of players are competent card counters, the cost-benefit analysis tips decidedly against shuffling. Nonetheless, the recent success of card counters has led some casinos to invest in automatic card shuffling machines, which also cost the casino money up front, but then allow for the cards to be shuffled without delaying the dealer.

that prevent the average counter from every really striking it rich:

- While an odds advantage of 2 percent is terrific for Las Vegas, viewed as a traditional investment, a 2 percent rate of return is no reason to give up your day job.
- Earning even that lousy 2 percent requires you to concentrate and lay off the booze and the dancers—heck, you might as well be in Salt Lake City.
- More significantly, earning that 2 percent rate of return requires you to bet in a predictable pattern—betting small amounts while the deck is cold (high cards already played on low cards remaining) and betting large amounts while the deck is hot (the opposite). Pit bosses and the "eye in the sky" are trained to recognize these patterns and throw you out of the casino.
- Gamblers are generally capital constrained. Even the best card counter will suffer losing streaks, no matter how good the odds become. But if you lose your whole stake, you're done. This is sometimes known as the "gambler's dilemma."

Well, there are problems, and then there are the geeks at MIT, who tend to solve problems. For years, teams of highly trained MIT students and dropouts prowled the casinos of Las Vegas and the floating casinos across America in the most organized and profitable card counting scheme ever mounted. (For a fun and quick read, pick up Ben Mezrich's *Bringing Down the House: The Inside Story of Six MIT Students Who Took Vegas for Millions.*) How did they do this?

- Interestingly, they did *not* develop a new, complex strategy for counting cards. Instead, they practiced the old strategy until they could implement it instinctually, without having to think about it. They did add some refinements, though, such as adjusting the regular "count"—or value of the deck—to reflect how much of the deck remained (as a +10 score means a lot more with twenty cards left than it does with one hundred cards left). This "true count" was more useful.

- They avoided the patterns of card counting through a strategy more reminiscent of Harvard Business School than the MIT Engineering Department: they worked in teams, and they specialized. To avoid predictable betting patterns, spotters would sit at one table and bet the same amount each time. Nothing suspicious there. When the deck got hot, they signaled what they called "Gorillas" or "Big Players" and through code transmitted the true count to them. Gorillas simply went from table to table, and always bet (and generally won) big. Nothing suspicious there, as casinos may notice *betting* patterns at *one table*, but don't notice *winning patterns* across *multiple tables*. Big Players were like Gorillas, but kept up the count once they sat down, improving the odds even further.
- They treated counting like a business, raising significant capital from investors. Back before money laundering was a major focus, they carried hundreds of thousands of dollars in cash to Las Vegas strapped to their clothes.

Using these methods, the MIT students earned consistent double-digit returns, and took in millions. That is, until one of their own betrayed them. But that's someone else's story. . . .

Could you duplicate the MIT students' feat? Probably not, even if you took the time and money. First, larger casinos are beginning to introduce continuous shuffling machines, which generally defeat counting. So you'll be plying your trade at backwaters where the comped rooms have cockroaches and the bacon at the gambler's breakfast is cold and greasy. Second, casinos are pioneering the use of facial recognition software, meaning that once you're spotted as a counter at one casino, you may be done at many. Third, and most importantly, you really don't want to spend that much time in a casino.

STARTING A SUCCESSFUL SMALL BUSINESS

Here's some cognitive dissonance for you: America's 22.4 million small businesses (those with fewer than five hundred employees) employ 51 percent of American workers, produce 51 percent of private sector output, and provide two thirds to three quarters of all new jobs. Yet, according to Small Business Administration estimates, one third of new businesses fail within two years; half fail within four years; and 60 percent fail within six years. How does so much of the U.S. economy depend (and depend successfully) on entrepreneurs pursuing an activity where the odds are stacked against them?

The Odds

Largely because, owing to efficient capital markets and forgiving bankruptcy laws, small business owners are able to bounce back, rapidly replacing old, failed businesses with new ones. According to Small Business Administration estimates, in 2001, of the 5.7 million businesses with more than one employee, about 585,800 were closed, but they were replaced by 574,500 new businesses. Furthermore, not all business closings are failures: 57 percent of business owners with one or more employees and 38 percent of sole proprietorships reported they were successful at closure. (Still, as we saw with penis size, there are always doubts when data are self-reported.)

Figure the odds against your small business surviving a respectable 6 years to be around 1.5 to 1.

Improving the Odds

Not all small businesses are created equal, and the chances of your succeeding will vary greatly depending on the choices you make. Herewith, some of the big decisions:

Choosing the right business. The odds of your small business succeeding will depend greatly on the kind of business you choose. Of course, there are limits to your choices. You can't start a deep sea fishing business in Kansas, and you may not want to open a dry cleaning business in West Virginia. That said, it pays to learn from the experience of others.

Odds of Succeeding at Small Business

Rank	Type of business	Number of businesses	% with net income	% with net losses
-	**All sole proprietorships (non-farm)**	**17,408,809**		
1	Surveying services	15,598	99%	1%
2	Water transport—contract pilots	4,700	98%	2%
3	Bus and limo transport—contract drivers	9,777	98%	2%
4	Optometrists	12,810	98%	2%
5	Other highway passenger transportation	21,637	96%	4%
6	Fuel oil dealers	3,573	95%	5%
7	Dentists	91,998	94%	6%
8	Textile mill products	5,668	94%	6%
9	Funeral services	9,272	94%	6%
10	Painting and paperhanging contractors	214,214	94%	6%
**and				
134	Hobby, toy, and game shops	25,936	46%	54%
135	Boat dealers	2,581	46%	54%
136	Food and kindred products	20,104	45%	55%
137	Used merchandise/antiques stores	103,373	44%	56%
138	Gift, novelty, and souvenir shops	105,035	41%	59%

Rank	Type of business	Number of businesses	% with net income	% with net losses
139	Women's ready-to-wear stores	20,019	41%	59%
140	Bookstores	10,338	40%	60%
141	Vending machine operations	45,797	37%	63%
142	Livestock breeding	26,188	25%	75%
143	Catalog and mail order operations	43,355	18%	82%

Source: BizMiner

So what do we learn? George Washington had it right in his choice of small business; surveyors have the best chance of succeeding. Your best odds, though, may lie with painting and paperhanging—not only is the success rate high, but also there is a lot of room for new entrants. The same probably cannot be said of flying a pontoon plane—a proven winner, to be sure, but it does help to have a pontoon plane and a large, placid body of water on which to land it—not to mention a pilot's license, a strong swimming stroke, and a tolerance for heights. But if your spouse should give you a call and say, "Honey, my uncle just died and—you're not going to believe this—he left us this pontoon plane he kept up in Anchorage," well, now you know it's *carpe pontoonum*.

At the other end of the spectrum, there are certainly small businesses to avoid if you wish to improve your odds of succeeding. Why would anyone choose one of these proven losers? Well, it looks like these are all businesses where you get to play with the merchandise—so, even if you do end up losing money, you've had a pretty good time. Games, boats, food, women's ready-to-wear shoes, books—all the major idle pursuits of the wealthy or retired are there for the money losing. (We are *not* going to discuss livestock breeding. . . .)

Use other people's money. Who can forget the classic *Other People's Money,* starring Danny DeVito and the lovely, redheaded Penelope Ann Miller, where DeVito portrays a corporate raider expert in leveraged buyouts—that is, borrowing money from other people to finance acquisitions of existing companies—and falls in love with the much younger, much taller daughter (Ms. Miller) in a family business he is targeting?

Actually, just about everyone except the most ardent, smitten Penelope Ann Miller fan has forgotten it, but the lesson for small business owners still holds true. Using OPM is the best way to start a small business, not only because you have less to lose, but because it gives your business greater credibility with suppliers and customers, and thereby increases your chances of success. While the best option is to attract equity from friends and neighbors—albeit friends and neighbors whose money (and, shortly thereafter, friendship) you wouldn't mind losing—the next best option is to approach the Small Business Administration and get some good old-fashioned government-sponsored loans.

Draft a business plan. You'd think this step would be an obvious one, but it is overlooked by many entrepreneurs. Studies show that the existence and quality of a business plan considerably improve the odds of success. You needn't look far for a guide to how to draft a business plan. Just go to the SBA's website, *www.sba.gov/starting/indexbusplans.html.*

Get advice. There is no shortage of resources ready to advise you on how to start a small business, and what to incorporate into your business plan. Amazon.com lists over 7,900 books on small business. And of course there's the Internet. The lawyers at *www.nolo.com*, for example, have handy tips on what type of business you should start—for example, "Choose a business that has a good chance of turning a profit." Want more trenchant advice like that? Just buy their Small Business Start-up Kit for $25.

PART
9

RANDOM GOOD FORTUNE

Your Whole World Waiting Behind Door Number Three

A whole generation grew up understanding the vagaries of random chance from Monty Hall and the wacky gang on *Let's Make a Deal*. The show initially ran from 1963 to 1977 and then reappeared intermittently until 1991. (A pale imitation, hosted by President Bush's nephew Billy, is now running on the Game Show Network.)

Hosted by the plaid-suited Canadian Monty Hall, and with prizes introduced by the miniskirted Carol Merrill, *Let's Make a Deal* was way ahead of its time. The next time you see Jay Leno wade into the *Tonight Show* audience ready to mix it up with his guests, think of Monty.* And Vanna White should be paying Carol Merrill a royalty.

For those too young to know (or too old to remember), *Let's Make a Deal* contestants were selected from the audience—heck, they played the game from their seats—based upon their ability to attract Monty's attention. They did so by wearing zany outfits and yelling, "Pick me! Pick me!!" Once picked, they were offered a choice of hidden prizes, say the small box Monty had sitting on a table, or a box on the showroom floor that was just the right size for a Maytag

* Some trivia: Monty Hall is the father of Joanna Gleason, whose name you certainly wouldn't recognize but whose face you certainly would, as she is a comedic character actress. Picture Leo McGarry's lawyer/girlfriend on *West Wing* or Dirk Diggler's mother in *Boogie Nights*.

refrigerator—or a llama. If the contestant initially won a decent prize, then Monty might give him or her a choice between (1) taking that prize and, say, $200 and going home, or (2) gambling on whatever was hidden behind a curtain or in some extremely large box. Sometimes the curtain or the box contained a wonderful prize—say a mink stole from Dicker and Dicker of Beverly Hills—but other times it contained a "zonk," like the aforementioned llama or a year's supply of Eskimo Pies or TurtleWax.*

The Odds

The climax of every show came, though, when the day's big winner got to go for the grand prize at the end of the show. The grand prize—say a brand-new Chevrolet—was hidden behind Door Number 1, Door Number 2, or Door Number 3, while the other two hid zonks or mediocre prizes. Obviously, your odds of winning the grand prize were 2 to 1. To build suspense, Monty invariably revealed what was behind one of the doors *not* chosen before moving on to the contestant's chosen door. (In addition to building suspense, this interim step was another chance to sell TurtleWax.) The opened door never contained the grand prize—that would have ruined the suspense—and instead was one of the two zonks.

Monty's simple act has yielded one of the great puzzles of mathematics, one that is now taught in schools and debated among academics. When highlighted by the "Ask Marilyn" column in *Parade* magazine, it prompted the greatest barrage of letters that the magazine had ever received.

The question is quite simple. You're playing the final round on *Let's Make a Deal* and have chosen your door. Sup-

* For more details on this and other shows, try *www.jumptheshark.com*.

pose that, after revealing that one of the other, non-chosen doors did *not* conceal the grand prize, Monty offers you the choice of switching doors. (For example, suppose you choose Door Number 1; Monty reveals that Door Number 2 hides a zonk; and Monty offers to let you switch to Door Number 3.) What is the appropriate response? (1) Stick with your original choice; (2) Switch to the remaining door; or (3) It doesn't matter. We're assuming here that Monty is not trying to trick you, and makes this offer in all cases, regardless of whether your initial choice was the grand prize or one of the two zonks. Think a minute before reading on.

Okay. Think of the problem this way. There will always be two doors that hold zonks, so regardless of whether you initially chose the grand prize or a zonk, Monty will always be able to show you a zonk not chosen. Your initial choice was either right or wrong, and the fact that he has shown you a door with a zonk behind it is neither surprising nor informative. It doesn't change the past. So switching doesn't change anything. Got it?

Well, actually, that's totally wrong. You should always switch, and your odds of winning the grand prize *double* if you do switch.

Innumerable explanations have been put forth to explain why the counterintuitive answer holds true. Many of them involve equations like "$P(?door1 \mid C1?R2) = 1-1/3 = 2/3$" or flow charts and diagrams. But consider three simpler ways to look at the problem. (For some reason, mathematicians have chosen to debate the problem using a car as the grand prize and a goat as the zonk, and we'll use that convention here.)

First, imagine that rather than three doors hiding two goats and one car, there are one thousand doors hiding 999 goats and one car. After your initial selection, 998 doors are opened to show goats. Do you still believe that the door you initially selected is equally as likely to hide the car as the other remaining door?

Second, consider the following syllogism:

- The chance that the car was put behind one of the other two doors is 2 in 3.
- If it was in fact put behind one of the other two doors and you switch, you are certain to get the car.
- Thus, you have a 2 in 3 chance of getting the car if you switch.

In other words, you're better off switching.

Finally, for the most stubborn and unconvinced, simply play the game a few dozen times. While playing out the game with doors or pieces of paper may take you a while, the Internet now offers various opportunities to do so electronically and quickly. Try, for example, the game at *www.stat.sc.edu/~west/javahtml/LetsMakeaDeal.html*. It shouldn't take you more than a minute or two to play a sufficient number of games to recognize that switching is a better strategy than sticking.

Now start making your friends and family miserable.

Guessing Heads or Tails

The tossing of a coin is the paradigm for even-money odds, the fairest way to decide which of two people will win some opportunity. But are the odds of getting, say, heads in the toss of a coin truly 1 to 1?

The Odds

The rather surprising answer is, "No, not exactly," and the story behind that answer is one of thousands of pale, emaciated statistics students, flipping and spinning coins at the direction of their pale, emaciated professors. But over the long term, their pathetic toiling may make you better off, so let's tip our beanies to them at the outset, and begin our crusade.

Statistics students have actually been studying the fairness of coins for some time. These studies consider three different ways of using a coin to make a decision: flipping, spinning, and tipping. Flipping is the traditional toss of the coin into the air, landing on the ground or the palm of your hand. Spinning involves placing the coin on the table and flicking it with a finger to make it spin. Tipping—too silly to consider here—involves standing the coin on its edge and seeing how it falls.

For a look at the major studies in this area, we turn to *Chance News*, the periodical bible of the statistical world. There we find a summary of some of the major U.S.penny-flipping experiments of our time:

Major Coin Flipping Experiments (U.S. penny)

Source	Type	Flips/Spins	Heads	% Heads	Std Deviation 95% conf.level
Buffon	Flip	4,040	2,048	50.70%	0.881 (.491, .522)
Pearson	Flip	24,000	12,012	50.05%	0.155 (.494, .507)
Kerrich	Flip	10,000	5,067	50.67%	1.34 (.477, .517)
Lock	Flip	29,015	14,709	50.70%	2.37 (.501, .513)
Peter	Spin	2,000	953	47.65%	-2.1 (.454,.499)
Dartmouth	Spin	2,000	913	45.65%	-3.89 (.434, .479)
Lock	Spin	20,422	9,197	45.00%	-14.19 (.443, .457)

Of course, "study" here is a bit of a misnomer, as we're not talking about scientists standing in a laboratory with goggles and lab coats on. Most of the studies are performed by undergraduates at bars. But a coin is a coin,* and a flat surface is a flat surface,** so there you are. As you'll see from the chart, flipping tends to favor heads, whereas spinning tends to favor tails. The flipping preference for heads is not statistically significant in any of the individual studies, but the persistent, though small, trend in that direction across all the studies seems probative. The preference for tails in the spinning experiments is far more significant.

Various theories have been put forth to explain why pennies tend to come up heads when flipped and tails when spun. The obvious candidates would be the relative weight and diameter of the head and tail sides of the coin, but, sadly, physics students have not taken up this cause with the enthusiasm of statistics students.

A major development in the coin flipping world came with the release of the Euro coin in 2002. Two Polish mathematicians, Tomasz Gliszczynski and Waclaw Zawadowski,

* Well, more or less. There has been some suggestion that the odds change as older coins, particularly pennies, have their surface worn down.

** Well, more or less. There has been some suggestion that tackier surfaces tend to diminish the tendency of spinning coins, discussed below, to come up tails.

suggested that the Euro had a strong bias in favor of heads when spun—albeit based only on a paltry 250 spins. (Such is the thirst for news by the 24/7 media that Gliszczynski and Zawadowski appeared in major media all over Europe, as well as National Public Radio in the States.) In an interview with the German newspaper *Die Welt*, Gliszczynski attributed the anomaly to the large image of King Albert II on the heads side of the coin. Subsequent research at Dartmouth appears to bear out this theory with respect to spinning. No statistically significant trend was noted with respect to flipping.

Euro Coin Experiment

Source	Type	Flips/Spins	Heads	% Heads
Gliszczynski/ Zawadowski	Spin	250	140	56.00%
Dartmouth	Spin	2000	1087	54.35%
Dartmouth	Flip	2000	994	49.70%

Improving the Odds

Now, if you're a man, you're probably wondering at this point, "Do these data have any relevance to the world of professional sports?" Of course, that's not how you'd phrase it—it would be more along the lines of "Hey, I gotta tell somebody at the Redskins—maybe they'd give me free tickets." But the answer is "Yes."

Obviously, to the extent that the referee at the beginning of a sporting event is flipping a standard U.S. coin, then the smart call for the athlete is tails. (Of course, this requires a team captain who can retain this information from the time he leaves the sideline to the time he makes the call, but the information can be written on wrist or jersey.) But the real potential lies in important games, where the leagues tend to use ceremonial coins—e.g., medals marking the fiftieth anniversary of the NFL, or the third consecutive British Premier league game without a fan fatality. Presumably, as these coins tend to be outsize affairs, they will tend to be weighted more off-center than most, and over time exhibit a general tendency toward either heads or tails.

Here, then, is one of the great tragedies of our time. No one appears to have taken it upon himself—it would never be a herself—to track the results of ceremonial coin flips. Gentlemen, start your TiVos.

LOOKING OVER A FOUR-LEAF CLOVER

Are you one of those people who used to find four-leaf clovers, or are you one of those people who used take a three-leaf clover, add one extra leaf removed from a nearby clover, and *pretend* to have found a four-leaf clover? If the latter (and, oh, the sad memories), how unlucky should you feel? What are the real odds of finding a four-leaf clover?

The Odds

White clover (*Trifolium repens*) is a member of the legume family, and its leaves generally have three leaflets.* Technically, *Trifolium repens* does not have three leaves, but rather one segmented leaf, with each of the segments referred to as a leaflet. By legend, the first leaflet brings hope; the second leaflet brings faith; the third leaflet brings love—and the fourth leaflet brings good luck. Of course, here good luck

* White clover is not to be mistaken for red clover (*Trifolium pratense*). Unlike white clover, which is green, red clover is actually red (or at least pink). Red clover is the state flower of Vermont, even though red clover does not grow in Vermont. If you think that's sad, consider that the four-leaf clover serves as the symbol of the Vermont lottery, even though red clover does not come in the four-leaf variety. Of course, that's truth in advertising, as your odds of winning the Vermont lottery are about the same as finding a four-leaf red clover. (The people of New Hampshire, who would ordinarily be quick to mock their New England rivals for such absurdities, are lying low in the wake of New Hampshire's "The Old Man in the Mountain"—the stone "face" that was the symbol of the state and featured on the New Hampshire quarter—tumbling into rubble.)

comes in the form of a genetic mutation. (Sort of like Carrot Top.)

According to—you guessed it—*www.fourleafclover.com,* the odds of picking up a good luck clover on the first try are about 10,000 to 1, or about the same as your dying in a fall. But those odds improve if you sit down in a clover patch and do a lot of patient searching.

Improving the Odds

Or cloning. Yes, cloning. Since 1998, the good folks at Yoke and Zoom in London (*www.yokeandzoom.com*) have bred four-leaf clover through cloning. Here, with the right plant to clone, the odds can be improved significantly. *Trifolium repens* clover is a low-growing perennial herb, which grows from the tip by sending out "runners" that take root. Although *Trifolium repens* naturally propagates through seed, Yoke and Zoom takes stem cuttings from the runners and grows them hydroponically (in water). This is commonly known as asexual reproduction or cloning.

Surprisingly, Yoke and Zoom is an organization devoted not to agricultural supplies or good luck curios but rather to supporting art through "installations," such as placing yellow boxes marked "Grim" and "Determination" at various sites around London. The installation *The Four Leaf Clover Cloning Station,* showing clover growing hydroponically, has been displayed around the United Kingdom.

To find out more, we contacted Nina Coulson, the cofounder at Yoke and Zoom. (The other cofounder is her husband, William. She is Zoom; he, Yoke.) Ms. Coulson's interest in *Trifolium repens* came about when she was renting a house while pursuing an art degree at the University of

Wales at Claereon. In the yard was a patch of clover that included, by her count, at least three hundred of the four-leaf variety. When the subsequent tenant declared that the clover was targeted for extinction (weed killer), Ms. Coulson returned to retrieve the plants.

Ms. Coulson has worked with Perry Michaelson Yeates at the Institute of Environment and Grassland Research in Wales. According to Mr. Yeates, hers is a rare find indeed. As noted, while one in ten thousand *Trifolium repens* will mutate to four-leaf form, some of these mutations are due in part to environmental, as opposed to genetic, factors, and that won't do for cloning purposes. Only one in one hundred thousand *Trifolium repens* will mutate in a fixed, natural environment. The odds of finding one that will mutate not just in its natural environment but also in a greenhouse or hydroponic environment—that is, one that mutates solely due to a hereditary, genetic defect and is thus cloning-portable—are 1,000,000 to 1. Such was Ms. Coulson's luck.

Taking cuttings from this super-mutated clover, Yoke and Zoom has been able to reduce the odds against finding a four-leaf clover to somewhere between 25 to 1 and 40 to 1. At peak growth times, the odds may drop to 4 to 1.

Can there be a dark side to four-leaf clover? Can the luck symbolized by that fourth leaf turn against you? The folks at Yoke and Zoom recognize the possibility in their installation *Rotten Luck*:

> Rotten luck is a conceptual piece. Set on a plinth, the work consists of nine petri dishes containing rotten, torn or insect bitten four leaf clovers. Rotten luck is the result of Yoke and Zoom's clovers that were diseased through a fungal infection in the plants or were partially consumed

> by insects. The work enables the audience to contemplate a physical manifestation of bad luck or misfortune, rather than simply displaying a four leaf clover as an innocent symbol or charm.

Please note that Yoke and Zoom is really not in the business of selling four-leaf clover. As Ms. Coulson explains, "We had a gallery offer to sell them for ten pounds each, but when I was picking them, I started thinking ten pounds, twenty pounds, thirty pounds," and that was not consistent with clover as art. Plus, according to legend, giving away clover brings good luck as well; selling them, not so much. So those who write or email asking for a sample are offered the chance to make a donation to offset Yoke and Zoom's costs, which generally works out to a couple of pounds. Through this compromise clover capitalism, the work moves forward.

We did not let Ms. Coulson go, though, without asking some of the tough questions. To wit: does she believe it is ethical to clone clover, or does this put us on a slippery slope to cloning ivy, rhododendrons, sheep, chimpanzees, and eventually humans? Ms. Coulson, who obviously had thought a lot about this issue, declared that she was strongly opposed to cloning when it involved genetic modification. Here, the cloning is organic, and does not involve any changes to the clover. Unfortunately, that answer seemed to make a lot of sense.

In closing, some of you may be wondering, "Hey, what kind of idiot actually believes that artists are growing *clover* hydroponically and selling it over the Internet. Any fool could have figured out that it's another green plant, of the three-leaf variety, that they're selling, and it's one that has been long prized by artists and rock musicians. And, by the way, what's that website again?" Well, although it turns out that marijuana has been grown hydroponically (a lot, actually), we can vouch for Yoke and Zoom. We have the four-leaf clover to prove it.

ENDNOTES

1. Bramlett, Matthew and William Mosher. "First marriage dissolution, divorce, and remarriage: United States," Advance Data From Vital and Health Statistics; No. 323. Hyattsville, MD: National Center for Health Statistics: 21.

2. Ibid.

3. Golden, Daniel. "Buying Your Way Into College." *Wall Street Journal,* March 12, 2003.

4. *www.pbs.org/wghb/masterpiece/archive/200/510e.html.*

5. For details on the Rhodes application process, see *www.rhodes-scholar.com.*

6. College ranks are according to *U.S. News & World Report.*

7. The National Safety Council's figures were based on 1999 data gathered from the National Center for Health Statistics and U.S. Census Bureau.

8. For more details on the exorcism process, see *www.themystica.com/mystica/articles/e/exorcism.html.*

9. *Cool, et al., and Blue Cross/Blue Sheild* v. *Olsen, et al.,* Circuit Ct., Outagamie Co., Wisconsin, Case No. 94 CV 707. The duck saga was also feature on CBS's *60 Minutes.*

10. Believe it or not, but NASA operates a website devoted entirely to the danger of the earth being destroyed by asteroids. If you'd like to worry yourself, look at *impact.arc.nasa.gov.*

11. *eMedicine Journal,* Volume 3, Number 6, June 26, 2002.

12. Boodman, Sandra. "No End to Errors: Three Years After a Landmark Report Found Pervasive Medical Mistakes in American Hospitals, Little Has Been Done to Reduce Death and Injury." *Washington Post,* December 3, 2001, HE01.

13. Mann, Charles C. "1491." *Atlantic Monthly* (March 2002).

14. Bailes, E., F. Gao, F. Bibollet-Ruche, V. Courgnaud, M. Peeters, P. A. Marx, B. H. Hahn, and P. M. Sharp. "Hybrid Origin of SIV in Chimpanzees." *Science,* 300 (June 6, 2003): 1713. Gao, F., E. Bailes, D. L. Robertson, Y. Chen, C. M. Rodenburg, S. F. Michael, L. B. Cummins, I. O. Arthur, M. Peeters, G. Shaw, P. M. Sharp, B. H. Hahn. "Origin of HIV-1 in the chimpanzee Pan troglodytes." *Nature,* 397 (1999): 436–440.

15. "Dating the Origin of the CCR5-32 AIDS-Resistance Allele by the Coalescence of Haplotypes." *American Journal of Human Genetics,* 62 (1998): 1507–1515.

16. Kleinfeld, J. "Could It Be a Big World After All? The 'Six Degrees of Separation' Myth." *Society* (2002).

17. *www.cnn.com/2002/HEALH/03/05/obesity.poll/*.

18. Leibel, R., M. Rosenbaum, and J. Hirsch. "Changes in Energy Expenditure Resulting from Altered Body Weight." *New England Journal of Medicine,* 332 (1995): 621–628.

19. Dimeglio, D. P., and Richard D. Mattes. "Liquid Versus Solid Carbohydrate: Effect on Food Intake and Body Weight." *International Journal of Obesity,* 794 (June 2000).

20. Wilcox, Allen J., M.D., Ph.D.; Clarice R. Weinberg, Ph.D.; and Donna D. Baird, Ph.D. "Timing of Sexual Intercourse in Relation to Ovulation—Effects on the Probability of Conception, Survival of the Pregnancy, and Sex of the Baby." *New England Journal of Medicine,* 333 (1995): 1517.

21. Witt, April, "False Confessions," *The Washington Post*, June 3–6 (2001) (four-day series).

ACKNOWLEDGMENTS

I have many people to thank for helping to put this book together. To begin with, all the friends (and some puzzled acquaintances) who have put up with my saying incessantly, "What are the odds?" or "Guess what I just found out about hemorrhoids." But there are some who have made much larger contributions.

Patrick Ferguson and Paul Beilke provided terrific research help and keen insights. Of course, there are tougher jobs in life than surfing the Internet looking for goofy statistics, but they did it with élan.* When Paul was sidelined by full-time employment and a new baby, Ted Wold picked up the standard and carried on. I am grateful to them all.

The dashing Brendan Cahill at Gotham Books provided a deft editorial touch, good humor, and quick turnarounds. I couldn't be more thankful. Thanks also to Lauren Marino, who gave *Life: The Odds* its first vote of confidence at Penguin Putnam, and Rick Willett, who provided an exceptional copyedit (including checking numerous sources to ensure that all supermodel names were spelled correctly).

My agent, Daniel Greenberg of Levine/Greenberg, has served as an advocate, trusted advisor, and friend. I would

* For those who think the Internet hasn't changed everything, it's interesting to note that I've never actually met Paul or Patrick. I was introduced to Paul by email, and found Patrick through an online listing at Georgetown. We corresponded by email, and I paid by PayPal. Truth be told, I'm not even 100 percent sure those are their real names. But perhaps we'll all meet up to celebrate publication.

advise anyone who has ever thought of writing a book—even if you aren't sure you have a good idea; even if you have only an outline on a couple of stray sheets of paper; even if you've been told over and over again that you have spelling and grammar "problems"—to call Daniel, night or day.

When it came time to ensure that the statistical foundation on which the book is built was solid, I relied on a trio of experts. Laurie Snell of Dartmouth College, while immeasurably overqualified for the task, agreed to review some key chapters. Matthew Parker and Alex Hanisch deftly spotted subtle errors and invalid assumptions; their skill and hard work allow me to sleep soundly at night.

Various friends suggested chapters, or helped on research, generally after drinking a fair amount of wine. Thanks to Allan Burrows, Mary Ann Frank, Chris Gallagher, Deb Jospin, Arnie Miller, and Andy Navarette.

Numerous folks provided helpful statistics or insights: the friendly women at the PGA Tour, the American Bowling Congress, Rick Breunig, Gibor Basri, James Levine, John Baker at Reed Business, Melody Lawrence and Corey Bray at the NCAA, and Michael Shackleford, the "Wizard of Odds."

Others gave the book a good read, and will now assume responsibility for all errors or misstatements. Direct your hate mail to Shirley Sagawa, David Voldemort (he who must not be named), Chris Bellini, and Eric Mogilnicki.

Special thanks to Dora Macia, Eugenia Romero, and others who helped out in countless ways.

This book couldn't have happened without the support of my family. Jack and Matt Baer are funny guys and boon companions who provided welcome distractions. Tommy Baer let me set up my iMac beside his bed, and hugged me while I typed.

And Shirley—well, for once, there are no words. . . .